E. Evangelisti (Ed.)

T0224093

Controllability and Observability

Lectures given at a Summer School of the
Centro Internazionale Matematico Estivo (C.I.M.E.),
held in Pontecchio (Bologna), Italy,
July 1-9, 1968

Springer

FONDAZIONE
CIME
ROBERTO CONTI

C.I.M.E. Foundation
c/o Dipartimento di Matematica "U. Dini"
Viale margagni n. 67/a
50134 Firenze
Italy
cime@math.unifi.it

ISBN 978-3-642-11062-7 e-ISBN: 978-3-642-11063-4
DOI:10.1007/978-3-642-11063-4
Springer Heidelberg Dordrecht London New York

Reprint of the 1st ed. C.I.M.E., Ed. Cremonese, Roma 1969
With kind permission of C.I.M.E.

Printed on acid-free paper

Springer.com

CENTRO INTERNAZIONALE MATEMATICO ESTIVO

(C.I.M.E.)

1° Ciclo - Sasso Marconi dal 1-9 Luglio 1968

CONTROLLABILITY AND OBSERVABILITY

Coordinatore: Prof. G. EVANGELISTI

CENTRO INTERNAZIONALE MATEMATICO ESTIVO
(C.I.M.E.)

LECTURES ON CONTROLLABILITY AND OBSERVABILITY

R.E. KALMAN
(Stanford-University)

Corso tenuto a Sasso Marconi (Bologna) dal 1 al 9 Luglio 1968

TABLE OF CONTENTS

R.E. Kalman

INTRODUCTION

The theory of controllability and observability has been developed, one might almost say reluctantly, in response to problems generated by technological science, especially in areas related to control, communication, and computers. It seems that the first conscious steps to formalize these matters as a separate area of (system-theoretic or mathematical) research were undertaken only as late as 1959, by KALMAN [1960b-c]. There have been, however, many scattered results before this time (see Section 12 for some historical comments and references), and one might confidently assert today that some of the main results have been discovered, more or less independently, in every country which has reached an advanced stage of "development" and it is certain that these same results will be rediscovered again in still more places as other countries progress on the road to development.

With the perspective afforded by ten years of happenings in this field, we ought not hesitate to make some guesses of the significance of what has been accomplished. I see two main trends:

(i) The use of the concepts of controllability and observability to study nonclassical questions in optimal control and optimal estimation theory, sometimes as basic hypotheses securing existence, more often as seemingly technical conditions which allow a sharper statement of results or shorter proofs.

(ii) Interaction between the concepts of controllability and observability and the study of structure of dynamical systems, such as: formulation and solution of the problem of realization, canonical forms, decomposition of systems.

The first of these topics is older and has been studied primarily from the point of view of analysis, although the basic lemma (2.7) is purely algebraic. The second group of topics may be viewed as "blowing up" the ideas inherent in the basic lemma (2.7), resulting in a more and more strictly algebraic point of view.

There is active research in both areas.

In the first, attention has shifted from the case of systems governed by finite-dimensional linear differential equations with constant coefficients (where success was quick and total) to systems governed by infinite-dimensional linear differential equations (delay differential equations, classical types of partial differential equations, etc.), to finite-dimensional linear differential equations with time-dependent coefficients, and finally to all sorts and subsorts of nonlinear differential equations. The first two topics are surveyed concurrently by WEISS [1969] while MARKUS [1965] looks at the nonlinear situation.

My own current interest lies in the second stream, and these lectures will deal primarily with it, after a rather hurried overview of the general problem and of the "classical" results.

Let us take a quick look at the most important of these "classical" results. For convenience I shall describe them in system-theoretic (rather than conventional pure mathematical) language. The mathematically trained reader should have no difficulty in converting them into his preferred framework, by digging a little into the references.

In area (i), the most important results are probably those which give more or less explicit and computable results for controllability and observability of certain specific classes of systems. Beyond these, there seem to be two main theorems:

THEOREM A. *A real, continuous-time, n-dimensional, constant, linear dynamical system* Σ *has the property "every set of* n *eigenvalues may be produced by suitable state feedback" if and only if* Σ *is completely controllable*.

The central special case is treated in great detail by KALMAN, FALB, and ARBIB [1969, Chapter 2, Theorem 5.10]; for a proof of the general case with background comments, refer to WONHAM [1967]. As a particular case, we have that every system satisfying the hypotheses of the theorem can be "stabilized" (made to have eigenvalues with negative real parts) via a suitable choice of feedback. This result is the "existence theorem" for algorithms used to construct control systems for the past three decades, and yet a conscious formulation of the problem and its mathematical solution go back to about 1963! (See Theorem D below.) The analogous problem for nonconstant linear systems (governed by linear differential equations with variable coefficients) is still not solved.

THEOREM B. ("Duality Principle") Every problem of controllability in a real, (continuous-time, or discrete-time), finite-dimensional, constant, linear dynamical system is equivalent to a controllability problem in a dual system.

This fact was first observed by KALMAN [1960a] in the solution of the optimal stochastic filtering problem for discrete-time systems, and was soon applied to several problems in system theory by KALMAN [1960b-c]. See also many related comments by KALMAN, FALB, and ARBIB [Chapters 2 and 6, 1969]. As a theorem, this principle is not yet known to be valid outside the linear area, but as an intuitive prescription it has been rather useful in guiding system-theoretic research. The problems involved here are those of fomulation rather than proof. The basic difficulties seem to point toward algebra and in particular category theory. System-theoretic duality, like the categoric one, is concerned with "reversing arrows". See Section 10 for a modern discussion of these points and a precise version of Theorem B.

Partly as a result of the questions raised by Theorem B and partly because of the algebraic techniques needed to prove Theorem A and related lemmas, attention in the early 1960's shifted toward certain problems of a structural nature which were, somewhat surprisingly at first, found to be related to controllability and observability. The main theorems again seem to be two:

THEOREM C. (Canonical Decomposition) Every real (continuous-time or discrete-time), finite-dimensional, constant, linear dynamical system may be canonically decomposed into four parts, of which only one part, that which is completely controllable and completely observable, is involved in the input/output behavior of the system.

The proof given by KALMAN [1962] applies to nonconstant systems only under the severe restriction that the dimensions of the subspace of all controllable and all unobservable states is constant on the whole real line. The result represented by Theorem C is far from definitive, however, since finite-dimensional linear, nonconstant systems admit at least four different canonical decompositions: it is possible and fruitful to dualize the notions of controllability and observability, thereby arriving at four properties, presently called

reachability and controllability

as well as

constructibility* and observability.

(See Section 2 definitions.) Any combination of a property from the first list with a property from the second list gives a canonical decomposition result analogous to Theorem C. The complexity of the situation was first revealed by WEISS and KALMAN [1965]; this paper contributed to a revival of interest (with hopes of success) in the special problems of nonconstant linear systems. Recent

----

*WEISS [1969] uses "determinability" instead of constructibility. The new terminology used in these lectures is not yet entirely standard.

progress is surveyed by WEISS [1969]. Intimately related to the canonical structure theorem, and in fact necessary to fully clarify the phrase "involved in the input/output behavior of the system", is the last basic result:

THEOREM D. (Uniqueness of Minimal Realization) Given the impulse-response matrix W of a real, continuous-time, finite-dimensional, linear dynamical system, there exists a real, continuous-time, finite-dimensional, linear dynamical system Σ_W which

(a) realizes W: that is, the impulse-response matrix of Σ_W is equal to W;

(b) has minimal dimension in the class of linear systems satisfying (a);

(c) is completely controllable and completely observable;

(d) is uniquely determined (modulo the choice of a basis at each t for its state space) by requirement (a) together with (b) or, independently, by (a) together with (c).

In short, for any W as described above, there is an "essentially unique" Σ_W of the same "type" which satisfies (a) through (c).

COROLLARY 1. If W comes from a constant system, there is a constant Σ_W which satisfies (a) through (c), and is uniquely determined by (a) + (b) or (a) + (c) (modulo a fixed choice of basis for its state space).

COROLLARY 2. All claims of Corollary 1 continue to hold if "impulse-response matrix of a constant, finite-dimensional system" is replaced by "transfer function matrix of a constant, finite-dimensional system".

The first general discussion of the situation with an equivalent statement of Theorem D is due to KALMAN [1963b, Theorems 7 and 8]. (This paper does not include complete proofs, or even an explicit statement of Corollaries 1 and 2, although they are implied by the general algorithm given in Section 7. An edited version of the original unpublished proof of Theorem D is given in KALMAN, FALB, and ARBIB [1969, Chapter 10, Appendix C].)

These results are of great importance in engineering system theory since they relate methods based on the Laplace transform (using the transfer function of the system) and the time-domain methods based on input/output data (the matrix W) to the state-variabl (dynamical system) methods developed in 1955-1960. In fact, by Corollary 1 it follows that the two methods must yield identical results; for instance, starting with a constant impulse-response matrix W, property (c) implies that the existence of a stable control lay is always assured by virtue of Theorem A. Thus it is only after the development represented by Theorems A-D that a rigorous justification is obtained for the intuitive design methods used in control engineering.

As with Theorem C, certain formulational difficulties arise in connection with a precise definition of a "nonconstant linear dynamical system". Thus, it seems preferable at present to replace in Theorem D "impulse-response matrix W" by "weighting pattern W" (or "abstract input/output map W") and "complete controllability" by "complete reachability". The definitive form of the 1963 theorem evolved through the works of WEISS and KALMAN [1965], YOULA [1966], and KALMAN; a precise formulation and modernized proof of Theorem D in the weighting pattern case was given recently by KALMAN, FALB, and ARBIB [1969, Chapter 10, Section 13.] A completely general discussion of what is meant by a "minimal realization" of a non-constant impulse-response matrix involves many technical complications due to the fact that such a minimal realization does not exist in the class of linear differential equations with "nice" coefficient functions. For the current status of this problem, consult especially DESOER and VARAIYA [1967], SILVERMAN and MEADOWS [1969], KALMAN, FALB, and ARBIB [1969, Chapter 10, Section 13] and WEISS [1969].

From the standpoint of the present lectures, by far the most interesting consequence of Theorem D is its influence, via efforts to arrive at a definitive proof of Corollary 1, on the development of the algebraic stream of system theory. The first proof of this important result (in the special case of distinct eigenvalues) is that of GILBERT [1963]. Immediately afterwards, a general proof was given by KALMAN [1963b, Section 7]. This proof, strictly computational and linear algebraic in nature, yields no theoretical insight although it is useful as the basis of a computer algorithm.

Using the classical theory of invariant factors, KALMAN [1965a] succeeded in showing that the solution of the minimal realization problem can be effectively reduced to the classical invariant-factor algorithm. This result is of great theoretical interest since it strongly suggests the now standard module theoretic approach, but it does not lead to a simple proof of Corollary 1 and is not a practical method of computation.

The best known proof of Corollary 1 was obtained in 1965 by B. L. Ho, with the aid of a remarkable algorithm, which is equally important from a theoretical and computational viewpoint. The early formulation of the algorithm was described by HO and KALMAN [1966], with later refinements discussed in HO and KALMAN [1969], KALMAN, FALB, and ARBIB [1969, Chapter 10, Section 11] and KALMAN [1969c]. Almost simultaneously with the work of B. L. Ho, the basic results were discovered independently also by YOULA and TISSI [1966] and by SILVERMAN [1966]. The subject goes back to the 19th century and centers around the theory of Hankel matrices; however, many of the results just referenced seem to be fundamentally new. This field is currently in a very active stage of development. We shall discuss the essential ideas involved in Sections 8-9. Many other topics, especially Silverman's generalization of the algorithm to nonconstant systems unfortunately cannot be covered due to lack of time.

Acknowledgment

It is a pleasure to thank C. I. M. E. and its organizers, especially Professors E. Bompiani, E. Sarti, and E. Belardinelli, for arranging a special conference on these topics. The sunny skies and hospitality of Italy, along with Bolognese food played a subsidiary but vital part in the success of this important gathering of scientists.

R. E. Kalman

1. CLASSICAL AND MODERN DYNAMICAL SYSTEMS

In mathematics the term dynamical system (synonyms: topological dynamics, flows, abstract dynamics, etc.) usually connotes the action of a one-parameter group T (the reals) on a set X, where X is at least a topological space (more often, a differentiable manifold) and the action is at least continuous. This setup is physically motivated, but in a very old-fashioned sense. A "dynamical system" as just defined is an idealization, generalization, and abstraction of Newton's world view of the Solar System as described via a finite set c nonlinear ordinary differential equations. These equations represent the positions and momenta of the planets regarded as point masses and are completely determined by the laws of gravitation, i.e., they do not contain any terms to account for "external" forces that may act on the system.

Interesting as this notation of a dynamical system may be (and is!) in pure mathematics, it is much too limited for the study of those dynamical systems which are of contemporary interest. There are at least three different ways in which the classical concept must be generalized:

(i) The time set of the system is not necessarily restricted to the reals;

(ii) A state $x \in X$ of the system is not merely acted upon by the "passage of time" but also by inputs which are or could be manipulated to bring about a desired type of behavior;

(iii) The states of the system cannot, in general, be observed. Rather, the physical behavior of the system is manifested through its outputs which are many-to-one functions of the state.

The generalization of the time set is of minor interest to us here. The notions of input and output, however, are exceedingly fundamental; in fact, controllability is related to the input and observability to the output. With respect to dynamical systems in the classical sense, neither controllability nor observability are meaningful concepts.

A much more detailed discussion of dynamical systems in the modern sense, together with rather detailed precise definitions, will be found in KALMAN, FALB, and ARBIB [1969, Chapter 1].

From here on, we will use the term "dynamical system" exclusively in the modern sense (we have already done so in the Introduction).

The following symbols will have a fixed meaning throughout the paper:

$$(1.1) \quad \begin{cases} T = \text{time set}, \\ U = \text{set of input values}, \\ X = \text{state set}, \\ Y = \text{set of output values}, \\ \Omega = \text{input functions}, \\ \varphi = \text{transition map}, \\ \eta = \text{readout map}. \end{cases}$$

The following assumptions will always apply (otherwise the sets above are arbitrary):

(1.2)

$$\begin{cases} T = \text{an ordered subset of the reals } \underline{\underline{R}}, \\ \Omega = \text{class of functions } T \rightarrow U \text{ such that} \\ \quad \text{(i) each function } \omega \text{ is undefined outside some} \\ \quad\quad \text{finite interval } J_\omega \subset T \text{ dependent on } \omega; \\ \quad \text{(ii) if } J_\omega \cap J_{\omega'} = \emptyset, \text{ there is a function} \\ \quad\quad \omega \in \Omega \text{ which agrees with } \omega \text{ on } J_\omega \text{ and} \\ \quad\quad \text{with } \omega' \text{ on } J_{\omega'}. \end{cases}$$

For most purposes later, T will be equal to $\underline{\underline{Z}}$ = (ordered) abelian group of integers; U, X, Y, Ω will be linear spaces; "undefined" can be replaced by "equal to 0"; and "functions undefined outside a finite interval" will mean the same as "finite sequences".

The most general notion of a dynamical system for our present needs is given by the following

(1.3) DEFINITION. *A dynamical system* Σ *is a composite object consisting of the maps* φ, η *defined on the sets* T, U, Ω, X, Y *(as above)*:

$$\varphi: \; T \times T \times X \times \Omega \rightarrow X,$$

$$: \; (t; \tau, x, \omega) \mapsto \varphi(t; \tau, x, \omega)$$

undefined whenever $t \geqq \tau$;

$$\eta: \; T \times X \rightarrow Y: \; (t, x) \mapsto \eta(t, x).$$

The transition map φ *satisfies the following assumptions:*

(1.4) $\varphi(t; t, x, \omega) = x;$

(1.5) $\varphi(t;\ \tau,\ x,\ \omega) = \varphi(t;\ s,\ \varphi(s;\ \tau,\ x,\ \omega),\ \omega);$

(1.6) <u>if</u> $\omega = \omega'$ <u>on</u> $[\tau,\ t]$, <u>then for all</u> $s \in [\tau,\ t]$

$\varphi(s;\ \tau,\ x,\ \omega) = \varphi(s;\ \tau,\ x,\ \omega').$

The definition of a dynamical system on this level of generality should be regarded only as a scaffolding for the terminology; interesting mathematics begins only after further hypotheses are made. For instance, it is usually necessary to endow the sets T, U, Ω, X, and Y with a topology and then require that φ and η be continuous.

(1.7) EXAMPLE. The classical setup in topological dynamics may be deduced from our Definition (1.3) in the following way. Let $T = \underline{\underline{R}}$ = reals, regarded as an abelian group under the usual addition and having the usual topology; let Ω consist only of the nowhere-defined function; let X be topological space; disregard Y and η entirely; define φ for <u>all</u> $t,\ \tau \in T$ and write it as

$$\varphi(t;\ \tau,\ x,\ \omega) = x\cdot(t - \tau),$$

that is, a function of x and $t - \tau$ alone. Check (1.4-5); in the new notation they become

$$x\cdot 0 = x \quad \text{and} \quad x\cdot(s + t) = (x\cdot s)\cdot t.$$

Finally, require that the map $(x,\ t) \mapsto x\cdot t$ be continuous.

(1.8) INTERPRETATION. The essential idea of Definition (1.3) is that it axiomatizes the notion of state. A dynamical system is informally a rule for state transitions (the function φ), together with suitable means of expressing the effect of the input on the state and the effect of the state on the output (the function η). The map φ is verbalized as follows: "an input ω, applied to the system Σ in state x at time τ produces the state $\varphi(t; \tau, x, \omega)$ at time t." The peculiar definition of an input function ω is used here mainly for technical convenience; by (1.6) only equivalence classes of inputs agreeing over $[\tau, t]$ enter into the determination of $\varphi(t; \tau, x, \omega)$. "$\omega$ not defined" at t means no input acts on Σ at time t.

The pair $(\tau, x) \in T \times X$ will be called an <u>event</u> of a dynamical system Σ.

In the sequel, we shall be concerned primarily with systems which are <u>finite-dimensional, linear, and continuous-time or discrete-time</u>. Often these systems will be also <u>real</u> and <u>constant</u> (= stationary or time-invariant). We leave the precise definition of these terms in the context of Definition (1.3) to the reader (consult KALMAN, FALB, or ARBIB [1969, Chapter 1] as needed) and proceed to make some ad hoc definitions without detailed explanation.

The following conventions will remain in force throughout the lectures whenever the linear case is discussed:

(1.9) <u>Continuous-time</u>. $T = \underline{\underline{R}}$, $U = \underline{\underline{R}}^m$, $X = \underline{\underline{R}}^n$, $Y = \underline{\underline{R}}^p$, Ω = all continuous functions $\underline{\underline{R}} \to \underline{\underline{R}}^m$ which vanish outside a finite interval.

(1.10) <u>Discrete-time</u>. $T = \underline{\underline{Z}}$, K = fixed field (arbitrary),

$U = K^m$, $X = K^n$, $Y = K^p$, Ω = all functions $\underline{\underline{Z}} \to K^m$ which are zero for all but a finite number of their arguments.

Now we have, finally,

(1.11) DEFINITION. A real, continuous-time, n-dimensional, linear dynamical system Σ is a triple of continuous matrix functions of time $(F(\cdot), G(\cdot), H(\cdot))$ where

$$F(\cdot): \underline{\underline{R}} \to \{n \times n \text{ matrices over } \underline{\underline{R}}\}$$

$$G(\cdot): \underline{\underline{R}} \to \{n \times m \text{ matrices over } \underline{\underline{R}}\},$$

$$H(\cdot): \underline{\underline{R}} \to \{p \times n \text{ matrices over } \underline{\underline{R}}\}.$$

These maps determine the equations of motion of Σ in the following manner:

$$(1.12) \quad \begin{cases} dx/dt = F(t)x + G(t)\omega(t), \\ y(t) = H(t)x(t), \end{cases}$$

where $t \in \underline{\underline{R}}$, $x \in \underline{\underline{R}}^n$, $\omega(t) \in \underline{\underline{R}}^m$, and $y(t) \in \underline{\underline{R}}^p$.

To check that (1.12) indeed makes Σ into a well-defined dynamical system in the sense of Definition (1.3), it is necessary to recall the basic facts about finite systems of ordinary linear differential equations with continuous coefficients. Define the map

$$\Phi_F(t, \tau): \underline{\underline{R}} \times \underline{\underline{R}} \to \{n \times n \text{ matrices over } \underline{\underline{R}}\}$$

to be the family of $n \times n$ matrix solutions of the linear differential equation

$$dx/dt = F(t)x, \quad x \in \underline{\underline{R}}$$

subject to the initial condition

$$\Phi_F(\tau, \tau) = I = \text{unit matrix}, \quad \tau \in \underline{\underline{R}}.$$

Then Φ_F is of class C^1 in both arguments. It is called the transition matrix of (the system Σ whose "infinitesimal" transition matrix is) $F(\cdot)$. From this standard result we get easily also the fact that the transition map of Σ is explicitly given by

(1.13) $$\varphi(t; \tau, x, \omega) = \Phi_F(t, \tau)x + \int_\tau^t \Phi_F(t, s)G(s)G'(s)\Phi_F'(t, s)ds$$

while the readout map is given by

(1.14) $$\eta(t, x) = H(t)x.$$

It is instructive to verify that φ indeed depends only on the equivalence class of ω's which agree on $[\tau, t]$.

In view of the classical terminology "linear differential equations with constant coefficients", we introduce the nonstandard

(1.15) DEFINITION. A real, continuous-time, finite-dimensional linear dynamical system $\Sigma = (F(\cdot), G(\cdot), H(\cdot))$ is called constant iff all three matrix functions are constant.

In strict analogy with (1.15), we say:

(1.16) DEFINITION. A discrete-time, finite-dimensional, linear, constant dynamical system Σ over K is a triple (F, G, H) of

$n \times n$, $n \times m$, $p \times n$ matrices over the field K. These maps determine the equations of motion of Σ in the following manner:

$$(1.17) \quad \begin{cases} x(t+1) = Fx(t) + G\omega(t), \\ y(t) = Hx(t), \end{cases}$$

where $t \in \mathbb{Z}$, $x \in K^n$, $\omega(t) \in K^m$, and $y(t) \in K^p$.

In the sequel, we shall use the notations $(F, G, -)$ or $F, -, H)$ to denote systems possessing certain properties which are true for any H or G.

Finally, we adopt the following convention, which is already implicit in the preceding discussion:

(1.18) DEFINITION. The dimension n of a dynamical system Σ is equal to the dimension of X_Σ as a vector space.

2. STANDARDIZATION OF DEFINITIONS AND "CLASSICAL" RESULTS

In this section, we shall be mainly interested in finite-dimensional linear dynamical systems, although the first two definitions will be quite general.

Let Σ be an arbitrary dynamical system as defined in Section 1. We assume the following slightly special property: There exists a state x^* and an input ω^* such that

$$\varphi(t;\ \tau,\ x^*,\ \omega^*) \ = \ x^* \quad \text{for all} \quad t,\ \tau \in T \quad \text{and} \quad t \geqq \tau.$$

For simplicity, we write x^* and ω^* as 0. (When X and Ω have additive structure, 0 will have the usual meaning.) The next two definitions refer to dynamical systems with this extra property.

(2.1) DEFINITION. An event (τ, x) is controllable iff[§] there exists a $t \in T$ and an $\omega \in \Omega$ (both t and ω may depend on (τ, x)) such that

$$\varphi(t;\ \tau,\ x,\ \omega) \ = \ 0$$

In words: an event is controllable iff it can be transferred to 0 in finite time by an appropriate choice of the input function ω. Think of the path from (τ, x) to $(t, 0)$ as the graph of a function defined over $[\tau, t]$.

[§]The technical word iff means if and only if.

Consider now a reflection of this graph about τ. This suggests a new definition which is a kind of "adjoint" of the definition of controllability:

(2.2) DEFINITION. An event (τ, x) is reachable iff there is an $s \in T$ and an $\omega \in \Omega$ (both s and ω may depend on (τ, x)) such that

$$x = \varphi(\tau; s, 0, \omega).$$

We emphasize: controllability and reachability are entirely different concepts. A striking example of this fact is encountered below in Proposition (4.26).

We shall now review briefly some well-known criteria for and relations between reachability and controllability in linear systems.

(2.3) PROPOSITION. In a real, continuous-time, finite-dimensional, linear dynamical system $\Sigma = (F(\cdot), G(\cdot), -\ )$, an event (τ, x) is

(a) reachable if and only if $x \in \text{range}\ \hat{W}(s, \tau)$ for some $s \in \underline{\underline{R}}$, $s < \tau$, where

$$\hat{W}(s, \tau) = \int_s^\tau \Phi_F(\tau, \sigma)G(\sigma)G'(\sigma)\Phi_F'(\tau, \sigma)d\sigma$$

(b) controllable if an only if $x \in \text{range}\ W(\tau, t)$ for some $t \in \underline{\underline{R}}$, $t > \tau$, where

$$W(\tau, t) = \int_\tau^t \Phi_F(\tau, s)G(s)G'(s)\Phi_F'(\tau, s)ds.$$

The original proof of (b) is in KALMAN [1960b]; both cases are treated in detail in KALMAN, FALB, and ARBIB [1969, Chapter 2,

Section 2]. Note that if $G(\cdot)$ is identically zero on $(-\infty, \tau)$ we cannot have reachability, and if $G(\cdot)$ is identically zero on $(\tau, +\infty)$ we cannot have controllability.

For a constant system, the integrals above depend only on the difference of the limits; hence, in particular

$$W(\tau, t) = \hat{W}(2\tau - t, \tau).$$

So we have

(2.4) PROPOSITION. <u>In a real, continuous-time, finite-dimensional, linear, constant dynamical system an event</u> (τ, x) <u>is reachable for all</u> τ <u>if and only if it is reachable for one</u> τ; <u>an event is reachable if and only if it is controllable.</u>

From (2.3) one can obtain in a straightforward fashion also the following much stronger result:

(2.5) THEOREM. <u>In a real, continuous-time, n-dimensional, linear, constant dynamical system</u> $\Sigma = (F, G, -)$ <u>a state</u> x <u>is reachable (or, equivalently, controllable) at any</u> $\tau \in \underline{\underline{R}}$ <u>if and only if</u>

$$x \in \text{span}\,(G, FG, \ldots) \subset \underline{\underline{R}}^n;$$

<u>if this condition is satisfied, we can choose</u> $s = \tau - \delta$, $t = \tau + \delta$, <u>with</u> $\delta > 0$ <u>arbitrary</u>. (The span of a sequence of matrices is to be interpreted as the vector space generated by the columns of these matrices.)

A proof of (2.5) may be found in KALMAN, HO, and NARENDRA [1963] and in KALMAN, FALB, and ARBIB [1969, Chapter 2, Section 3]. A trivial but noteworthy consequence is the fact that the definition of reachable states of Σ is "coördinate-free":

(2.6) COROLLARY. *The set of reachable (or controllable) states of* Σ *in Theorem* (2.5) *is a subspace of the real vector space* X_Σ, *the state space of* Σ.

Very often the attention to individual states is unnecessary and therefore many authors prefer to use the terminology "Σ is completely reachable at τ" for "every event (τ, x), τ = fixed, $x \in X_\Sigma$, is reachable", or "Σ completely reachable" for "every event in Σ is reachable", etc. Thus (2.5), together with the Cayley-Hamilton theorem, implies the

(2.7) BASIC LEMMA. *A real, continuous-time, n-dimensional, linear, constant dynamical system* $\Sigma = (F, G, -)$ *is completely reachable if an only if*

(2.8) $$\text{rank}\,(G, FG, \ldots, F^{n-1}G) \;=\; n.$$

Condition (2.8) is very well-known; it or equivalent forms of it have been discovered, explicitly used, or implicitly assumed by many authors. A trivially equivalent form of (2.7) is given by

(2.9) COROLLARY 1. *A constant system* $\Sigma = (F, G, -)$ *is completely reachable if and only if the smallest* F-*invariant subspace of* X_Σ *containing (all column vectors of)* G *is* X_Σ *itself.*

A useful variant of the last fact is given by

(2.10) COROLLARY 2. (W. Hahn) A constant system $\Sigma = (F, G, -)$ is completely reachable if and only if there is no nonzero eigenvector of F which is orthogonal to (every column vector of) G.

Finally, let us note that, far from being a technical condition, (2.5) has a direct system-theoretic interpretation, as follows:

(2.11) PROPOSITION. The state space X_Σ of a real, continuous-time, n-dimensional, linear, constant dynamical system $\Sigma = (F, G, -)$ may be written as a direct sum

$$X_\Sigma = X_1 \oplus X_2,$$

which induces a decomposition of the equations of motion as (obvious notations)

$$(2.12)\quad \begin{cases} dx_1/dt = F_{11}x_1 + F_{12}x_2 + G_1u(t), \\ dx_2/dt = F_{22}x_2. \end{cases}$$

The subsystem $\Sigma_1 = (F_{11}, G_1, -)$ is completely reachable. Hence a state $x = (x_1, x_2) \in X_\Sigma$ is reachable if and only if $x_2 = 0$.

PROOF. We define X_1 to be the set of reachable states of Σ; by (2.5) this is an F-invariant subspace of X_Σ. Hence, by finite-dimensionality, X_1 is a direct summand in X_Σ. By construction, every state in X_1 is reachable, and (every column vector of)

G belongs to X_1. The F-invariance of X_1 implies that $F_{11} = 0$, which implies the asserted form of the equations of motion. □

(2.13) REMARK. Note that X_2 is not intrinsically defined (it depends on an arbitrary choice in completing the direct sum). Hence to say that "$(0, x_2)$ is an unreachable (or uncontrollable) state if $x_2 \neq 0$" is an abuse of language. More precisely: the set of all reachable (or controllable) states has the structure of a vector space, but the set of all unreachable (or uncontrollable) states does not have such structure. This fact is important to bear in mind for the algebraic development which follows after this section and also in the definition of observability and constructibility below. In general, the direct sum cannot be chosen in such a way that $F_{12} = 0$.

While condition (2.8) has been frequently used as a technical requirement in the solution of various optimal control problems in the late 1950 s, it was only in 1959-60 that the relation between (2.8) and system theoretic questions was clarified by KALMAN [1960b-c] via Definition (2.2) and Propositions (2.5) and (2.11). (See Section 11 for further details.) In other words, without the preceding discussion the use of (2.8) may appear to be artificial, but in fact it is not, at least in problems in which control enters, because, by (2.12) control problems stated for X_Σ are nontrivial only with respect to the intrinsic subspace X_1.

The hypothesis "constant" is by no means essential for Proposition (2.11), but we must forego further comments here.

For later purposes, we state some facts here for discrete-time, constant linear systems analogous to those already developed for their continuous-time counterparts. The proofs are straightforward and therefore omitted (or given later, for illustrative purposes).

(2.14) PROPOSITION. A state x of a real, discrete-time, n-dimensional, linear, constant dynamical system $\Sigma = (F, G, -)$ is reachable if and only if

(2.15) $x \in \text{span}\,(G, FG, \ldots, F^{n-1}G)$.

Thus such a system is completely reachable if and only if (2.8) holds.

(2.16) PROPOSITION. A state x of the system Σ described in Proposition (2.14) is controllable if and only if

(2.17) $x \in \text{span}\,(F^{-1}G, \ldots, F^{-n}G)$,

where

$$F^{-k}G = \{x\colon\ F^k x = g_i,\ g_i = \text{column vector of } G\}.$$

(2.18) PROPOSITION. In a real, discrete-time, finite-dimensional, linear, constant dynamical system $\Sigma = (F, G, -)$ a reachable state is always controllable and the converse is always true whenever $\det F \neq 0$.

Note also that Propositions (2.11) and its proof continue to be correct, without any modification, when "continuous-time" is replaced by "discrete-time".

Now we turn to a discussion of observability.

The original definition of observability by KALMAN [1960b, Definition (5.23)] was concocted in such a way as to take advantage of vector-space duality. The conceptual problems surrounding duality are easy to handle in the linear case but are still by no means fully understood in the nonlinear case (see Section 10). In order to get at the main facts quickly, we shall consider here only the linear case and even then we shall use the underlying idea of vector-space duality in a rather ad-hoc fashion. The reader wishing to do so can easily turn our remarks into a strictly dual treatment of facts (2.1)-(2.12) with the aid of the setup introduced in Section 10.

(2.19) DEFINITION. An event (τ, x) in a real, continuous-time, finite-dimensional, linear dynamical system $\Sigma = (F(\cdot), -, H(\cdot))$ is unobservable iff

$$H(s)\Phi_F(s, \tau)x = 0 \quad \text{for all} \quad s \in [\tau, \infty).$$

(2.20) DEFINITION. With respect to the same system, an event (τ, x) is unconstructible* iff

----

*In the older literature, starting with KALMAN [1960b, Definition (5.23)], it is this concept which is called "observability". By hindsight, the present choice of words seems to be more natural to the writer.

$$H(\sigma)\Phi_F(\sigma,\ \tau)x \;=\; 0 \quad \underline{\text{for all}} \quad \sigma \in (-\infty,\ \tau].$$

The motivation for the first definition is obvious: the "occurrence" of an unobservable event cannot be detected by looking at the output of the system after time τ. (The definition subsumes $\omega = 0$, but this is no loss of generality because of linearity.) The motivation for the second definition is less obvious but is in fact strongly suggested by statistical filtering theory (see Section 10). In any case, Definition (2.21) complements Definition (2.20) in exactly the same way as Definition (2.1) complements Definition (2.2).

From these definitions, it is very easy to deduce the following criteria:

(2.21) PROPOSITION. <u>In a real, continuous-time, finite-dimensional, linear dynamical system</u> $\Sigma = (F(\cdot),\ -,\ H(\cdot))$ <u>an event</u> $(\tau,\ x)$ <u>is</u>

(a) <u>unobservable if and only if</u> $x \in \text{kernel } \hat{M}(\tau,\ t)$ <u>for all</u> $t \in \underline{\underline{R}},\ t > \tau,$ <u>where</u>

$$\hat{M}(\tau,\ t) \;=\; \int_\tau^t \Phi_F'(s,\ \tau)H'(s)H(s)\Phi_F(s,\ \tau)ds;$$

(b) <u>unconstructible if and only if</u> $x \in \text{kernel } M(s,\ \tau)$ <u>for all</u> $s \in \underline{\underline{R}},\ s < \tau,$ <u>where</u>

$$M(s,\ \tau) \;=\; \int_s^\tau \Phi_F'(\sigma,\ \tau)H'(\sigma)H(\sigma)\Phi_F(\sigma,\ \tau)d\sigma.$$

PROOF. Part (a) follows immediately from the observation: $x \in$ kernel $M(\tau, t) \Leftrightarrow H(s)\Phi_F(s, \tau)x = 0$ for all $s \in [\tau, t]$. Part (b) follows by an analogous argument. □

(2.22) REMARK. Let us compare this result with Proposition (2.3), and let us indulge (only temporarily) in abuses of language of the following sort:*

$$(\tau, x) = \underline{\text{unreachable}} \Leftrightarrow x \in \text{kernel } \hat{W}(\tau, t)$$
$$\underline{\text{for all}} \quad t > \tau$$

and

$$(\tau, x) = \underline{\text{observable}} \Leftrightarrow x \in \text{range } \hat{M}(\tau, t)$$
$$\underline{\text{for some}} \quad t > \tau.$$

From these relations we can easily deduce the so-called "duality rules"; that is, problems involving observability (or constructibility) are converted into problems involving reachability (or controllability) in a suitably defined dual system. See KALMAN, FALB, and ARBIB [1969, Chapter 2, Proposition (6.12)] and the broader discussion in Section 10.

We <u>will</u> say, by slight abuse of language, that a system is <u>completely observable</u> whenever 0 is the only unobservable state. Thus the Basic Lemma (2.7) "dualizes" to the

(2.23) PROPOSITION. <u>A real, continuous-time or discrete-time, n-dimensional, linear, constant dynamical system</u> $\Sigma = (F, -, H)$

----

*All this would be strictly correct if we agreed to replace "direct sum" in Proposition (2.11) and its counterpart (2.25) by "orthogonal direct sum"; but this would be an arbitrary convention which, while convenient, has no natural system-theoretic justification. Reread Remark (2.13).

is completely observable if and only if

(2.24) $\quad \operatorname{rank}\,(H', F'H', \ldots, (F')^{n-1}H') = n.$

By duality, complete constructibility in a continuous-time system is equivalent to observability; in a discrete-time system this is not true in general but it is true when $\det F \neq 0$.

It is easy to see also that (2.11) "dualizes" to:

(2.25) PROPOSITION. *The state space* X_Σ *of a real, continuous-time or discrete-time, n-dimensional, linear, constant dynamical system* $\Sigma = (F, -, H)$ *may be written as a direct sum*

$$X_\Sigma = X_1 \oplus X_2$$

and the equations of Σ *are decomposed correspondingly as*

$$\begin{aligned} dx_1/dt &= F_{11}x_1, \\ dx_2/dt &= F_{21}x_1 + F_{22}x_2, \\ y(t) &= H_2x_2(t). \end{aligned}$$

PROOF. Proceed dually to the proof of Proposition (2.11), beginning with the definition of X_1 as the set of all *un*observable states of Σ. □

Combining Propositions (2.11) and (2.25) gives Theorem C as in KALMAN [1962].

This completes our survey of the "classical" results related to reachability, controllability, observability, and constructibility.

The remaining lectures will be concerned exclusively with discrete-time systems. The main motivation for the succeeding developments will be the algebraic criteria (2.8) and (2.24) as well as a deeper examination of Theorems C and D of the Introduction.

3. DEFINITION OF STATES VIA NERODE EQUIVALENCE CLASSES

A classical dynamical system is essentially the action of the time set T (= reals) on the states X. In other words, the states are acted on by an abelian group, namely ($\underline{\underline{R}}$ + usual definition of addition). This is a trivial fact, but it has deep consequences. A (modern) dynamical system is the action of the inputs Ω on X; in exact analogy with the classical case, to the abelian structure on T there corresponds an (associative but noncommutative) semigroup structure on Ω. The idea that Ω always admits such a structure was apparently overlooked until the late 1950's when it became fashionable in automata theory (school of SCHUTZENBERGER). This seems to be the "right" way of translating the intuitive notion of dynamics into mathematics, and it will be fundamental in our succeeding investigations.

It is convenient to assume from now on, until the end of these lectures, that

(3.1) T = time set = $\underline{\underline{Z}}$ = additive (ordered) group of integers.

Since we shall be only interested in constant systems from here on, we shall adopt the following normalization convention:*

----

*In the discrete-time nonconstant case, we would have to deal with $\underline{\underline{Z}}$ copies of Ω, each normalized with respect to a different particular value of $\tau \in \underline{\underline{Z}}$.

(3.2) <u>No element of</u> Ω <u>is defined for</u> $t > \tau = 0$.

In view of (3.2), we can define the "length" $|\omega|$ of ω by

$$|\omega| = \max\{-t \in \underline{\underline{Z}}: \ \omega \text{ is not defined for any } s < t\}.$$

Before defining the semigroup on Ω, we introduce another fundamental notion of dynamics: the (left) shift operator σ_Ω, defined for all $q \geq 0$ in $\underline{\underline{Z}}$ by

(3.3) $$\sigma_\Omega^q: \ \Omega \to \Omega: \ \omega \mapsto \sigma_\Omega^q \omega: \ t \to \omega(t + q).$$

Note that the definition of σ_Ω is compatible with the normalization (3.2).

If $J_\omega \cap J_{\omega'}$ = empty for $\omega, \omega' \in \Omega$, we define the <u>join</u> of ω and ω' as the function

(3.4) $$\omega \vee \omega' = \begin{cases} \omega & \text{on } J_\omega, \\ \omega' & \text{on } J_{\omega'}. \end{cases}$$

When Ω has an additive structure, then we replace $\omega \vee \omega'$ by $\omega + \omega'$.

(3.5) DEFINITION. <u>There is an associative operation</u> $\circ: \ \Omega \times \Omega \to \Omega$, <u>called concatenation, defined by</u>

$$\circ: \ (\omega, \nu) \mapsto \sigma_\Omega^{|\nu|}\omega \vee \nu.$$

Note that, by (3.2) through (3.4), $\circ$ is well defined.

Note also that the asserted existence of concatenation rests on the fact that Ω is made up of functions defined over <u>finite</u> intervals in T. We might express the content of (3.5) also as: Ω is a semigroup with valuation, since evidently $|\omega \circ \nu| = |\omega| + |\nu|$.

In view of (3.5), it is natural to use an abbreviated notation* also for the transition function, as follows:

$$(3.6) \qquad x \circ \omega = \varphi(0; -|\omega|, x, \omega)$$

Now we come to an important nonclassical concept in dynamical systems, whose evolution was strongly influenced by problems in communications and automata theory: a <u>discrete-time constant input/output map</u>

$$(3.7) \qquad f: \Omega \to Y: \omega \mapsto f(\omega) = y(1)$$

We interpret this map as follows: $y(1)$ is the output of some system Σ (say, a digital computer) when Σ is subjected to the (finite) input sequence ω, assuming that Σ is some fixed initial equilibrium state before the application of ω. This definition automatically incorporates the notions of "discrete-time" as well as "causal" or "dynamics" (the latter because $y(t)$ is <u>not defined</u> for $t < 1$). However, (3.7) does not clearly imply "constancy" (implicitly, however, this is clear from the normalization assumption (3.2) on Ω). To make the definition more forceful, we extend f to the map

$$(3.8) \qquad \bar{f}: \Omega \to \Gamma = Y \times Y \ \ldots \ \text{(infinite cartesian product)}$$
$$: \omega \mapsto (f(\omega), f(\sigma_\Omega\omega), \ldots) = (y(1), y(2), \ldots).$$

Interpretation: $\bar{f}$ gives the output sequence $\gamma = (y(1), y(2), \ldots$ of the system Σ after $t = 0$ resulting from the application of an

----

*Observe that $x \circ \omega$ is the strict analog of the notation xt customary in topological dynamics. The action of ω on x satisfies $x \circ (\omega \circ \nu) = (x \circ \omega) \circ \nu$ in view of (1.5).

input ω which stops at $t = 0$.

This definition expresses causality more forcefully and incorporates constancy, provided we define the (left) shift operator σ_Γ on Γ so as to be compatible with (3.3). So, for any $\tau \geq 0$, $\tau \in \mathbb{Z}$, let

$$(3.9) \qquad \sigma_\Gamma\colon \Sigma \to \Gamma\colon \gamma \mapsto \sigma_\Gamma\gamma\colon t \mapsto \gamma(t + \tau)$$
$$:(y(1), y(2), \ldots) \mapsto (y(\tau + 1), y(\tau + 2), \ldots)$$

Note: the operator σ_Ω "appends" an undefined term at 0, the operator σ_Γ "discards" the term $y(1)$.

Now, dropping the bar over f, we adopt

(3.10) DEFINITION. A discrete-time, constant input/output map (of some underlying dynamical system Σ) is any map f such that the following diagram

$$\begin{array}{ccc} \Omega & \xrightarrow{f} & \Gamma \\ \sigma_\Omega \downarrow & & \downarrow \sigma_\Gamma \\ \Omega & \xrightarrow[f]{} & \Omega \end{array}$$

is commutative. We say that f is linear iff it is a K-vector space homomorphism.

It will be convenient to regard (3.10) as the external definition of a dynamical system, in contrast to the internal definition set up in Section 1.

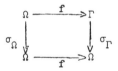

Intuitively, we should think of f as a highly idealized kind of experimental data; namely, f incorporates all possible information that could be gained by subjecting the underlying system to experiments in which only input/output data is available. This point of view is related to experimental physics the same way as the classical notion of a dynamical system is related to Newtonian (axiomatic) physics.

The basic question which motivates much of what will follow can now be formulated as follows:

(3.11) PROBLEM OF REALIZATION. Given only the knowledge of f (but of course also of $\underline{\underline{Z}}$, Ω, and Γ) how can we discover, in a mathematically consistent, rigorous, and natural way, the properties of the system Σ which is supposed to underlie the given input/output map f?

This suggests immediately the following fundamental concept:

(3.12) DEFINITION. A fixed dynamical system Σ (internal definition, as in Section 1) is a realization of a fixed input/output map f_o iff $f_o = f_{\Sigma_o}$, that is, f_o is identical with the input/output map of Σ_o.

In view of the notations of Section 1 plus the special convention (3.6), the explicit form of the realization condition is simply that

$$(3.13) \qquad f_o(\omega) = \eta_{\Sigma_o}(\varphi_{\Sigma_o}(0; -|\omega|, *, \omega))$$

for all ω Ω. The symbol $*$ stands for an arbitrary equilibrium state in which Σ_o remains, by definition, until the application of ω. (Later we simply take $*$ to be 0.)

To solve the realization problem, the critical step is to induce a definition of X (of some Σ_0) from the given f_0. It is rather surprising that this step turns out to be trivial, on the abstract level. (On the concrete level, however, there are many unsolved problems in actually computing what X is. In Section 8, we shall solve this problem, too, but only in the linear case.) The essential idea seems to have been published first by NERODE [1958]:

(3.14) DEFINITION. Make the concatenation semigroup Ω into a monoid by adjoining a neutral element $\emptyset$ (which is the nowhere-defined function on $\underline{\underline{Z}}$). Then $\omega \equiv_f \omega'$ (read: ω is Nerode equivalent to ω' with respect to f) iff

$$f(\omega \circ \nu) \;=\; f(\omega' \circ \nu) \quad \text{for all} \quad \nu \in \Omega.$$

There are many intuitive, physical, historical, and technical reasons (which are scattered throughout the literature and concentrated especially strongly in KALMAN, FALB, and ARBIB [1969]) for using this as the

(3.15) MAIN DEFINITION. The set of equivalence classes under $\equiv_f$, denoted as $X_f = \{(\omega)_f\colon \omega \in \Omega\}$, is the state set of the input/output map f.

Let us verify immediately that (3.15) makes mathematical sense:

(3.16) PROPOSITION. <u>For each linear, constant input/output map</u> f <u>there exists a dynamical system</u> Σ_f <u>such that</u>

(a) Σ_f <u>realizes</u> f;

(b) $X_{\Sigma_f} = X_f$.

PROOF. We show how to induce Σ_f, given f. We define the state set of Σ_f by (b). Further, we define the transition function of Σ_f by

$$(3.17) \qquad x \circ \nu = (\omega)_f \circ \nu \triangleq (\omega \circ \nu)_f \quad \text{for all} \quad \nu \in \Omega,\ x \in X_f.$$

We must check that $\circ$ on the left of $\triangleq$ is well defined (note two different uses of $\circ$!), that is, independent of the representation of x as $(\omega)_f$. This follows trivially from (3.14). Now we define the <u>readout map</u> of Σ_f by

$$(3.18) \qquad \eta_{\Sigma_f}: X_f \to Y: (\omega)_f \mapsto f(\omega)(1)$$

Again, this map is well defined since we can take $\nu = \emptyset$ as a special case in (3.14). Then

$$\eta_{\Sigma_f}(x \circ \nu) = \eta_{\Sigma_f}((\omega \circ \nu)_f) = f(\omega \circ \nu),$$

and the realization condition (3.6) is verified. Hence claim (a) is correct. □

(3.19) COMMENTS. In automata theory, Σ_f is known as the <u>reduced form</u> of any system which realizes f. Clearly, any two reduced forms are isomorphic, in the set-theoretic sense, since the set X_f is intrinsically defined by f. (This observation is a weak version of Theorem D of the Introduction; here "uniqueness" means "modulo a permutation of the labels of elements in the set X_f".) Notice also that Σ_f is completely reachable since, by Definition (3.15), every element $x = (\omega)_f$ of X_f is reachable via any element ω' in the Nerode equivalence class $(\omega)_f$. As to observability of Σ_f, see Section 10.

4. MODULES INDUCED BY LINEAR INPUT/OUTPUT MAPS

We are now ready to embark on the main topics of these lectures. It is assumed that the reader is conversant with modern algebra (especially: abelian groups, commutative rings, fields, modules, the ring of polynomials in one variable, and the theory of elementary divisors), on the level of, say, VAN DER WAERDEN, LANG [1965], HU [1965] or ZARISKI and SAMUEL [1958, Vol. 1]. The material covered from here on dates from 1965 or later.

Standing assumptions until Section 10:

(4.1) All systems $\Sigma = (F, G, H)$ are discrete-time, linear, constant, defined over a fixed field K (but not necessarily finite-dimensional).

Our immediate objective is to provide the setup and proof for the

(4.2) FUNDAMENTAL THEOREM OF LINEAR SYSTEM THEORY. The natural state set X_f associated with a discrete-time, linear, constant input-output map f over a fixed field K admits the structure of a finitely generated module over the ring $K[z]$ of polynomials (with indeterminate z and coefficients in K).

(4.3) COMMENTS. Since the ring $K[z]$ will be seen to be related to the inputs to Σ, this result has a superficial resemblance to the fact that in an arbitrary dynamical system Σ the state set X_Σ admits the action of a semigroup, namely Ω_Σ (see (3.6) and related footnote). It turns out, however, that this action of Ω on X, which results from combining the concatenation product in Ω with the definition of states via Nerode equivalence, is incompatible with the additive structure of Ω [KALMAN, 1967, Section 3]. Our theorem asserts the existence of an entirely different kind of structure of X. This structure, that of a K[z]-module, is not just a consequence of dynamics, but depends critically on the additive structure on Ω and on the linearity of f. The relevant multiplication is not (noncommutative) concatenation but (commutative) convolution (because convolution is the natural product in K[z]); dynamics is thereby restated in such a way that the tools of commutative algebra become applicable. In a certain rather definite sense (see also Remark (4.30)), Theorem (4.2) expresses the algebraic content of the method of the Laplace transformation, especially as regards the practices developed in electrical engineering in the U.S. during the 1950's.

The proof of Theorem (4.2) consists in a long sequence of canonical constructions and the verification that everything is well defined and works as needed.

In view of (4.1) and the conventions made in Section 1, Ω may be viewed as a K-vector space and $\omega(t) = 0$ for almost all $t \in \underline{\underline{Z}}$ and all $\omega \in \Omega$. By convention (3.2), we have assumed also that $\omega(t) = 0$ for all $t > 0$. As a result, we have that:

(a) $\Omega \approx K^m[z]$ as a K-vector space. Let us exhibit the isomorphism explicitly as follows:

$$(4.4) \qquad \omega \approx \sum_{t \in \underline{\underline{Z}}} \omega(t) z^{-t} \in K^m[z].$$

By (3.2), the sum in (4.4) is always finite. The isomorphism obviously preserves the K-linear structure on Ω. In the sequel, we shall not distinguish sharply between ω as a function $T \to K^m$ and ω as an m-vector polynomial.

(b) Ω <u>is a free</u> $K[z]$-<u>module with</u> m <u>generators, that is,</u> $\Omega \approx K^m[z]$ <u>also in the</u> $K[z]$-<u>module sense</u>. In fact, we define the action of $K[z]$ on Ω by scalar multiplication as

$$\cdot : \; K[z] \times \Omega \to \Omega : \; (\pi, \omega) \mapsto \pi \cdot \omega$$

where

$$(4.5) \qquad \pi \cdot \omega = \begin{bmatrix} \pi\omega_1 \\ \vdots \\ \pi\omega_m \end{bmatrix} \qquad (\omega_j \in K[z], \; j = 1, \ldots, m).$$

The product of π with the components of the vector ω is the product in $K[z]$. We write the scalar product on the left, to avoid any confusion with notation (3.6). It is easy to see that the module axioms are verified; Ω is obviously free, with generators

$$(4.6) \qquad e_j = \begin{bmatrix} 0 \\ \vdots \\ 1 \\ \vdots \\ 0 \end{bmatrix} \leftarrow \text{j-th position}, \; j = 1, \ldots, m.$$

(c) <u>On</u> Ω <u>the action of the shift operator</u> σ_Ω <u>is represented by multiplication by</u> z. This, of course, is the main reason for introducing the isomorphism (4.4) in the first place.

(d) <u>Each element of</u> Γ <u>is a formal power series in</u> z^{-1}. In fact, (4.4) suggests viewing z^t as an abstract representation of $-t \in \underline{\underline{Z}}$; hence we define

(4.7) $$\gamma \approx \sum_{t \in \underline{\underline{Z}}} \gamma(t) z^{-t} \in K^p[[z^{-1}]].$$

By (3.8) and (4.1), $\gamma(t) \in K^p$ for each $t > 1$ and is zero (or not defined) for $t < 1$. In general the sum is taken over infinitely many nonzero terms; there is no question of convergence and the right-hand side of (4.7) is to be interpreted stictly algebraically as a formal power series. Since $\gamma(0)$ is always zero (see (3.8)), we can say also that

(e) Γ <u>is isomorphic to the</u> K-<u>vector subspace of</u> $K^p[[z^{-1}]]$ <u>(formal power series in</u> z^{-1} <u>with coefficients in</u> K^p) <u>consisting of all power series with</u> 0 <u>first term</u>.

The first nontrivial construction is the following:

(f) Γ <u>has the structure of a</u> $K[z]$ <u>module, with scalar multiplication defined as</u>

(4.8) $$\cdot : K[z] \times \Gamma \to \Gamma: \quad (\pi, \gamma) \mapsto \pi \cdot \gamma = \pi(\sigma_\Gamma)\gamma.$$

This product may be interpreted as the ordinary product of a power series in z^{-1} by a polynomial in z, followed by the deletion of all terms containing no negative powers of z. The verification of the module axioms is straightforward.

(g) f <u>is a</u> $K[z]$ <u>homomorphism</u>. This is an immediate consequence of the fact that f = constant (see (3.10)) and that multiplication by z corresponds to the left shift operators on Ω and Γ.

(h) <u>The Nerode equivalence classes of</u> f <u>are isomorphic with</u> Ω/kernel f. This is an easy but highly nontrivial lemma, connecting Nerode equivalence with the module structure on Ω. The proof is an immediate consequence of the formula

$$(4.9) \qquad \omega \circ \nu = z^{|\nu|}\omega + \nu.$$

In fact, by K-linearity of f, (4.9) implies

$$f(\omega \circ \nu) = f(\omega' \circ \nu) \quad \text{for all} \quad \nu \in \Omega$$

if and only if

$$f(z^k \cdot \omega) = f(z^k \cdot \omega') \quad \text{for all} \quad k \geq 0 \quad \text{in} \quad \underline{\underline{Z}}.$$

The proof of Theorem (4.2) is now complete, since the last lemma identifies X_f as defined by (3.15) with the $K[z]$ quotient module Ω/kernel f.

We write elements of the latter as $[\omega]_f = \omega$ + kernel f; then it is clear that X_f as a $K[z]$-module is generated by $[e_1]_f, \ldots, [e_m]_f$, since Ω itself is generated by $e_1, \ldots, e_m$ (see (4.6)). Note also that the scalar product in Ω/kernel f is

$$(4.10) \qquad (\pi, [\omega]_f) \mapsto \pi \cdot [\omega]_f = [\pi \cdot \omega]_f.$$

The last product above (that in Ω) has already been defined in (4.5). The reader should verify directly that (4.10) gives a well-defined scalar product.

(4.11) REMARK. There is a strict duality in the setup used to define f. From the point of view of homological algebra [MAC LANE 1963], this duality looks as follows. Since every free module is projective, <u>the natural map</u>

$$\mu: \Omega \to X_f: \omega \mapsto [\omega]_f$$

<u>exhibits</u> X_f <u>as the image of a projective module</u>. On the other hand, there is a bijection between the set X_f and the set

$$\Xi_f = f(\Omega) \subset \Gamma.$$

Ξ_f is clearly a $K[z]$-submodule of Γ (with $z \cdot f(\omega) = f(z \cdot \omega)$), and so X_f and Ξ_f are isomorphic also as $K[z]$-modules. It is known that Γ is an injective module [MAC LANE 1963, page 95, Exercise 2] So <u>the natural map</u> $X_f \to \Xi_f$: $[\omega]_f \mapsto f(\omega)$ <u>exhibits</u> X_f <u>as a submodule of an injective module</u>. This fact is basic in the construction of the "transfer function" associated with f (Section 7), but its full implications are not yet understood at present.

There is an easy counterpart of Theorem (4.2) which concerns a dynamical system given in "internal" form:

(4.12) PROPOSITION. <u>The state set</u> X_Σ <u>of every discrete-time, finite-dimensional, linear, constant dynamical system</u> $\Sigma = (F, G, -)$ <u>admits the structure of a</u> $K[z]$-<u>module</u>.

PROOF. By definition (see (1.10)), $X = K^n$ is already a K-vector space. We make it into a $K[z]$-module by defining

(4.13) $\quad \cdot : K[z] \times K^n \to K^n : (\pi, x) \mapsto \pi(F)x.$ □

(4.14) COMMENT. The construction used in the proof of (4.12) is the classical trick of studying the properties of a fixed linear map $F: K^n \to K^n$ via the $K[z]$-module structure that F induces on K^n by (4.13). In view of the canonical construction of Σ_f provided by Proposition (3.16), the state set X can be treated as a $K[z]$-module irrespective as to whether X is constructed from f $(X = X_f)$ or given a priori as part of the specification of Σ $(X = X_\Sigma)$. Thus the $K[z]$-module structure on X is a nice way of uniting the "external" and the "internal" definitions of a dynamical system. Henceforth we shall talk about a (discrete-time, linear, constant dynamical) system Σ somewhat imprecisely via properties of its associated $K[z]$-module X_Σ.

We shall now give some examples of using module-theoretic language to express standard facts encountered before.

(4.15) PROPOSITION. _If_ X _is the state-module of_ Σ, _the map_ F_Σ _is given by_ $X \to X: x \mapsto z \cdot x$.

PROOF. This is obvious from (4.13) if $X = X_\Sigma$. If $X = X_f = X_{\Sigma_f}$, then we find that, by (1.17),

$$\begin{aligned} x(1) &= Fx(0) + G\omega(0), \\ &= F[\xi]_f + G\omega(0); \end{aligned}$$

since $x(0)$ results from input ξ, $x(1)$ results from input $z \cdot \xi + \omega(0)$

and we get

$$= [z\cdot\xi + \omega(0)]_f,$$
$$= z\cdot[\xi]_f + [\omega(0)]_f,$$
$$= z\cdot[\xi]_f + G\omega(0).$$

So the assertion is again verified. □

Now we can replace Proposition (2.14) by the much more elegant

(4.16) PROPOSITION. _A system_ $\Sigma = (F, G, -)$ _is completely reachable if and only if the columns of_ G _generate_ X_Σ.

PROOF. The claim is that complete reachability is equivalent to the fact that every element $x \in X_\Sigma$ is expressible as

$$x = \sum_{j=1}^{m} \pi_j g_j, \quad \pi_j \in K[z], \quad G = [g_1, \ldots, g_m].$$

In view of (4.15), this is the same as requiring that x be expressible as

$$x = \sum_{j=1}^{m} \pi_j(F) g_j;$$

this last condition is equivalent to complete reachability by (2.14). □

(4.17) COROLLARY. _The reachable states of_ Σ _are precisely those of the submodule of_ X_Σ _generated by (the columns of)_ G.

(4.18) REMARK. The statement that "Σ is not completely reachable" simply means that X is _not_ generated by those vectors which make up the matrix G in the specification of the input side of the system Σ.

It does not follow that X cannot be finitely generated by some <u>other</u> vectors. In fact, to avoid unnecessary generality, we shall henceforth assume that

X <u>is always finitely generated over</u> $K[z]$.

From the system-theoretic point of view, the case when we need infinitely many generators, that is, infinitely many input channels, seems rather bizzare at present.

(4.19) PROPOSITION. <u>The system</u> X_f <u>is completely reachable</u>.

PROOF. Obvious from the notation: a state $x = [\xi]_f$ is reached by $\xi \in \Omega$. □

(4.20) PROPOSITION. <u>The system</u> X_f <u>is completely observable</u>.

PROOF. Obvious from Lemma (h) above: $\eta([\omega]_f) = f(\omega) = 0$ iff $\omega \in [0]_f$, which says that the only unobservable state of X_f is $0 \in X_f$. □

Let us generalize the last result to obtain a module-theoretic criterion for complete observability. There are two technically different ways of doing this. The first depends on the observation that the "dual" of a submodule (see Corollary (4.17)) is a quotient module. The second defines observability via the "dual" system $(F', H', -)$ associated with $(F, -, H)$.

Consider a dynamical system $\Sigma = (F, -, H)$ and the corresponding $K[z]$-module X_Σ and K-homomorphism $H: X_\Sigma \to Y = K^p$. We can extend H

to a $K[z]$-homomorphism $\bar{H}$ (look back at (?.8)) by setting

$$\bar{H}: X_\Sigma \to \Gamma$$
$$x \mapsto (Hx,\ H(z\cdot x),\ H(z^2\cdot x),\ \dots\ ).$$

From Definition (2.19) we see that no nonzero element of the quotient module $X_\Sigma/\text{kernel } \bar{H}$ is unobservable. Hence, by abuse of language, we can say that $X_\Sigma/\text{kernel } \bar{H}$ is the module of observable states of Σ. Thus we arrive at phrasing the counterparts of (4.16-17) in the following language:

(4.21) PROPOSITION. _A system_ $\Sigma = (F, -, H)$ _is completely observable if and only if the quotient module_ $X_\Sigma/\text{kernel } \bar{H}$ _is isomorphic with_ X_Σ.

(4.22) COROLLARY. _The observable states of_ Σ _are to be identified with the elements of the quotient module_ $X_\Sigma/\text{kernel } \bar{H}$.

(4.23) TERMINOLOGY. The preceding considerations suggest viewing a system Σ as essentially the same "thing" as a module X. Strictly speaking, however, knowing $\Sigma = (F, G, H)$ gives us not only $X_\Sigma = X_F$ (see (4.13)) but also a quotient module X_Σ^o (over kernel $\bar{H}$) of a submodule (that generated by G) of X_F, that is

$$X_\Sigma^o = K[z]G/\text{kernel } \bar{H}.$$

If $X_\Sigma^o \approx X_\Sigma$ we say that X_Σ is _canonical_ (relative to the given G, H).

To be more precise, let us observe the following stronger version of (4.19-20):

(4.24) CORRESPONDENCE THEOREM. <u>There is a bijective correspondence between</u> $K[z]$-<u>homomorphisms</u> $f: \Omega \to \Gamma$ <u>and the equivalence class of completely reachable and completely observable systems</u> Σ <u>modulo a basis change in</u> X_Σ.

Detailed discussion of this result is postponed until Section 7.

A stricter observation of the "duality principle" leads to

(4.25) DEFINITION. <u>The</u> K-<u>linear dual of</u> $\Sigma = (F, G, H)$ <u>is</u> $\Sigma^* = (F', H', G')$ (' = <u>matrix transposition</u>). <u>The states of</u> Σ^* <u>are called costates of</u> Σ.

The following fact is an immediate consequence of this definition:

(4.26) PROPOSITION. <u>The state set</u> X_{Σ^*} <u>of</u> Σ^* <u>may be given the structure of</u> $K[z^{-1}]$ <u>module, as follows</u>: (i) <u>as a vector space</u> X_{Σ^*} <u>is the dual of</u> X_Σ <u>regarded as a</u> K-<u>vector space</u>, (ii) <u>the scalar product in</u> X_{Σ^*} <u>is defined by</u>

$$(z^{-1} \cdot x^*)(x) = x^*(Fx).$$

(4.26A) REMARK. We <u>cannot</u> define X_{Σ^*} as $\mathrm{Hom}_{K[z]}(X_\Sigma, K[z])$ equal to $K[z]$-linear dual of X_Σ, because every torsion module M over an integral domain D has a trivial D-dual. However, the reader can verify (using the ideas to be developed in Section 6) that X_{Σ^*} defined above is isomorphic with $\mathrm{Hom}_{K[z]}(X_\Sigma, K(z)/K[z])$. See BOURBAKI [<u>Algèbre</u>, Chapter 7 (2e éd.), Section 4, No. 8].

Now we verify easily the following dual statements of (4.16-17):

(4.27) PROPOSITION. A system $\Sigma = (F, -, H)$ is completely observable if and only if H' generates $X_{\Sigma*}$.

(4.28) COROLLARY. The observable COstates of $\Sigma*$ are precisely the reachable states of $\Sigma*$, that is, those of the submodule of $X_{\Sigma*}$ generated by H'.

We have eliminated the abuse of language incurred by talking about "observable states" through introduction of the new notion of "observable COstates". The full explication of why this is necessary (as well as natural) is postponed until Section 10.

The preceding simple facts depend only on the notion of a module and are immediate once we recognize the fact that F may be eliminated from statements such as (2.8) by passing to the module induced by F via (4.13). But module theory yields many other, less obvious results as well, which derive mainly from the fact that $K[z]$ is a principal-ideal domain.

We recall: an element m of an R-module M (R = arbitrary commutative ring) has torsion iff there is a $r \in R$ such that $r \cdot m = 0$. If this is not the case, m is free. Similarly, M is said to be a torsion module iff every element of M has torsion. M is a free module if no nonzero element has torsion. If $L \subset M$ is any subset of M, the annihilator A_L of L is the set

$$A_L = \{r: r \cdot \ell = 0 \text{ for all } \ell \in L\};$$

it follows immediately that A_L is an ideal in R. Note also that the statement "M is a torsion module" does not imply in general that A_L is nontrivial, that is, $A_L \neq 0$. (Counterexample: take an M which is not finitely generated.)

Coupling these notions with the special fact that, for us, $R = K[z]$, we get a number of interesting system-theoretic results:

(4.29) PROPOSITION. Σ _is finite-dimensional if and only if_ X_Σ _is a torsion_ $K[z]$-_module_.

COROLLARY. _If_ X_Σ _is free,_ Σ _is infinite dimensional_.

PROOF. We recall that "Σ = finite-dimensional" is defined to be "X_Σ = finite-dimensional as a K-vector space". See (1.18).

Sufficiency. By assumption X is finitely generated by, say, q nonzero elements $x_1, \ldots, x_q$ of X_Σ (which are not necessarily the columns of G). Hence

$$A_X = A_{x_1} \cap \ldots \cap A_{x_q}$$

Since $K[z]$ is a principal-ideal domain, each of the A_{x_j} is a principal ideal, say, $\gamma_j K[z]$ with $\gamma_j \in K[z]$. If X_Σ is a torsion module, then $\deg \gamma_j = n_j > 0$ for all $j = 1, \ldots, q$. For otherwise γ_j is either zero (and then x_j is free, which is a contradiction) or a unit which implies $x_j = 0$ contr y to assumption. Hence we can replace each expression

$$x = \sum_{j=1}^{q} \pi_j \cdot x_j, \quad \pi_j \in K[z]$$

by the simpler one

$$x = \sum_{j=1}^{q} [\pi_j \ (\mathrm{mod}\ \gamma_j)] \cdot x_j,$$

which shows that X_Σ, as a K-module, is generated by the finite set

$$x_1, \ z \cdot x_1, \ \ldots, \ z^{n_1 - 1} \cdot x_1, \ x_2, \ \ldots, \ x_q.$$

Necessity. Let ψ_F be the minimal polynomial of the map $F: x \mapsto z \cdot x$. If X_Σ is finite-dimensional as a K-module, $\deg \psi_F > 0$. This means (by the usual definition of the minimal polynomial in matrix theory or more generally in linear algebra) that ψ_F annihilates every $x \in X_\Sigma$ so that X_Σ is a torsion $K[z]$-module. □

Notice, from the second half of the proof, that the notion of a minimal polynomial can be extended from K-linear algebra to $K[z]$-modules. In fact, the same argument gives us also the well-known

(4.30) PROPOSITION. Every finitely generated torsion module M over a principal-ideal domain R has a nontrivial minimal p ynomial ψ_M given by $A_M = \psi_M R$.

(4.31) COROLLARY. If a $K[z]$-module X is finitely generated with q generators and minimal polynomial ψ_X, then

$$\dim X \ \text{(as K-vector space)} \ \leqq \ q \cdot \deg \psi_X.$$

(4.32) REMARK. The fact that Σ_f is completely reachable and is therefore generated by m vectors allows us to estimate the dimension of Σ_f by (4.31) knowing only $\deg \psi_{X_f}$ but without having computed

X_f itself. (Knowing X_f explicitly means knowing $F: x \mapsto z \cdot x$, etc.) In other words, the module-theoretic setup considerably enhances the content of Proposition (3.16). Guided by these observations, we shall develop in Section 8 explicit algorithms for calculating $\dim \Sigma_f$ directly from f without first having to compute F.

(4.33) PROPOSITION. *If X_Σ is a free $K[z]$-module, no state of Σ can be simultaneously reachable and controllable.*

PROOF. We recall that "X_Σ = free" means that X_Σ is (isomorphic to) a finite sum of copies of $K[z]$. Suppose for simplicity that $X_\Sigma = K[z]$. Then x = reachable means that $x = \xi \cdot 1$ for some $\xi \in K[z]$. Similarly, x = controllable means that $z^{|\omega|} \cdot x + \omega \cdot 1 = 0$ for some $\omega \in K[z]$. Hence if x has both properties,

$$(z^{|\omega|}\xi + \omega) \cdot 1 = (\xi \circ \omega) \cdot 1 = 0.$$

This shows that 1 is annihilated by $\xi \circ \omega$, the input ξ followed by ω, which contradicts the assumption that X_Σ is free. □

The most important consequence of Theorem (4.2) is due to the fact that through it we can apply to linear dynamical systems the well-known

(4.34) FUNDAMENTAL STRUCTURE THEOREM FOR FINITELY GENERATED MODULES OVER A PRINCIPAL IDEAL DOMAIN R (Invariant Factor Theorem for Modules). *Every such module M with m generators is isomorphic to*

(4.35) $$R/\psi_1 R \oplus \ldots \oplus R/\psi_r R \oplus R^s$$

where the $R/\psi_i R$ *are quotient rings of* R *viewed as modules over* R, *the* ψ_i *(called the invariant factors of* M*) are uniquely determined by* M *up to units in* R, $\psi_i | \psi_{i-1}$, $i = 2, \ldots, q$, *and, as usual,* R^s *denotes the free* R*-module with* s *generators; finally,* $r + s \leqq m$.

Various proofs of this theorem are referenced in KALMAN, FALB, and ARBIB [1969, page 270], and one is given later in Section 6.

Note: The divisibility conditions imply that M *is a torsion module iff* $s = 0$ and then $\psi_M = \psi_1$.

One important consequence of this theorem (others in Section 7) is that it gives us the most general situation when X_Σ is not a torsion module Σ. For instance, combining (4.33) with (4.34), we get

(4.36) PROPOSITION. *A system cannot be simultaneously completely reachable and completely controllable if its* $K[z]$*-module* X *has any* ∞*-dimensional components (i.e.,* $s > 0$ *in* (4.35)).

(4.37) REMARK. Although our entire development in this section may be regarded as a deep examination of Proposition (2.14), most of our comments apply equally well to (2.7), since both statements rest on the *same* algebraic condition (2.8). In fact, the only remaining thing to be "algebraized" is the notion of "continuous-time". We shall not do this here. Once this last step is taken, the algebraization of the Laplace transform (as related to *ordinary* linear differential equations) will be complete.

5. CYCLICITY AND RELATED QUESTIONS

We recall that an R-module M (R = arbitrary ring) is <u>cyclic</u> iff there is an element $m \in M$ such that $M = Rm$. [It would be better to say that such a module is monogenic: generated by one element m.]

If M is cyclic, the map $R \to M: r \mapsto r \cdot m$ is an epimorphism and has kernel A_m, the annihilating ideal of m. This plus the homomorphism theorem gives the well-known

(5.1) PROPOSITION. <u>Every cyclic R-module M with generator m is isomorphic with the quotient ring R/A_m viewed as an R-module.</u>

This result is much more interesting when, as in our case, R is not only commutative and a principal-ideal domain, but specifically the polynomial ring $K[z]$.

So let X be a cyclic $K[z]$-module with generator g and let $A_g = \psi_g K[z]$, where ψ_g is the <u>minimal</u> or <u>annihilating</u> polynomial of g. By commutativity and cyclicity, $A_g = A_X$. Hence ψ_g is a minimal polynomial also for X. Write $\psi_g = \psi_X = \psi$. In view of (5.1), $X \approx K[z]/\psi K[z]$. Let us recall some features of the ring $K[z]/\psi K[z]$:

(i) Its elements are the residue classes of polynomials $\pi \pmod{\psi}$, $\pi \in K[z]$. Write these as $[\pi]$ or $[\pi]_\psi$. Multiplication is defined as $[\pi]\cdot[\sigma] = [\pi\sigma]$.

(ii) Each $[\pi]$ is either a unit or a divisor of zero. In fact, $[\pi]$ is a unit iff (π, ψ) = greatest common divisor of π, ψ is a unit in $K[z]$ (that is, $(\pi, \psi) \in K$). Then

$$\sigma\pi + \tau\psi = 1 \quad (\sigma, \tau \in K[z])$$

so that $[\sigma]$ is the inverse of $[\pi]$. On the other hand, if $(\pi, \psi) = \theta \neq$ unit in $K[z]$, then both $[\pi]$ and $[\psi/\theta]$ are zero divisors since $[\pi]\cdot[\psi/\theta] = [(\pi/\theta)\psi] = 0$.

(iii) If ψ is a prime in $K[z]$ (that is, an irreducible polynomial with respect to coefficients over the ground field K), then by (ii) $K[z]/\psi K[z]$ is a field. This is a very standard construction in algebraic number theory.

Since it is awkward to compute with equivalence classes $[\pi]$, we shall often prefer to work with the standard representative of $[\pi]$, namely a polynomial $\tilde{\pi}$ of least degree in $[\pi]$. $\tilde{\pi}$ is uniquely determined by $[\pi]$ and the condition $\deg \tilde{\pi} < \deg \psi$. Henceforth $\tilde{}$ will always be used in this sense.

The next two assertions are immediate:

(5.2) PROPOSITION. $K[z]/\psi K[z]$ _as a_ K-_vector space is isomorphic to the_ K-_vector space_ $\mathbb{P}^{(n)} = \{\tilde{\xi} \in K[z]: \deg \tilde{\xi} < n = \deg \psi\}$. $K[z]/\psi K[z]$ _is also isomorphic to_ $\mathbb{P}^{(n)}$ _as a_ $K[z]$-_module, provided we define the scalar product in_ $\mathbb{P}^{(n)}$ _by_ $(\pi\cdot\tilde{\xi}) \mapsto \widetilde{\pi\xi}$.

(5.3) PROPOSITION. _If_ X_Σ _is cyclic with minimal polynomial_ ψ, _then_ $\dim \Sigma = \deg \psi$.

Looking back at Theorem (4.34), we see that the most general $K[z]$-module is a direct sum of cyclic $K[z]$-modules. By combining (5.3) and (4.34) and using the fact that dimension is additive under direct summing, we can replace (4.31) by the following exact result:

(5.4) PROPOSITION. *If* X_Σ *is a torsion module with invariant factors* $\psi_1, \ldots, \psi_q$ *then*

$$\dim \Sigma = \deg \psi_1 + \ldots + \deg \psi_q.$$

A simple but highly useful consequence of cyclicity is the so-called *control canonical form* [KALMAN, FALB, and ARBIB, 1969, page 44] for a completely reachable pair (F, g) where g is an $n \times 1$ matrix. We shall now proceed to deduce this result.

Observe first that "(F, g) completely reachable" is equivalent to "g generates X_F, the module induced by F via (4.13)." Let

$$\begin{aligned} \chi_F(z) &= \det (zI - F), \\ &= z^n + \alpha_1 z^{n-1} + \ldots + \alpha_n, \quad \alpha_1 \in K; \end{aligned}$$

then χ_F is the characteristic (and also the) minimal polynomial for X_F. [This is a well-known fact of module theory. See for example KALMAN, FALB, and ARBIB [1969, Chapter 10, Section 7] for detailed discussion.] As in KALMAN [1962], consider the vectors

$$(5.5)\quad \begin{cases} e_n = g = 1\cdot g = \chi_F^{(1)}(z)\cdot g, \\ e_{n-1} = z\cdot g + \alpha_1\cdot g = \chi_F^{(2)}(z)\cdot g, \\ \quad\vdots \\ e_1 = z^{n-1}\cdot g + a_1 z^{n-2}\cdot g + \dots + \alpha_{n-1}\cdot g = \chi_F^{(n)}(z)\cdot g \end{cases}$$

in X_F. [For consistency, $\chi_F^{(n+1)}(z) = \chi_F(z)$.] These vectors are easily seen to be linearly independent over K. They generate X_F since $X_F \approx \mathbb{P}^{(n)}$ as a K-vector space (Proposition (5.2)). Hence $e_1, \dots, e_n$ are a basis for X_F as a K-vector space. With respect to this basis, the K-homomorphism

$$z\colon K^n \to K^n\colon\ x \mapsto z\cdot x$$

is represented by the matrix

$$(5.6)\quad F = \begin{bmatrix} 0 & 1 & 0 & \dots & 0 & 0 \\ 0 & 0 & 1 & \dots & 0 & 0 \\ \dots & \dots & \dots & & \dots & \dots \\ 0 & 0 & 0 & \dots & 0 & 1 \\ -\alpha_r & -\alpha_{n-1} & -\alpha_{n-2} & \dots & -\alpha_2 & -\alpha_1 \end{bmatrix}$$

[This is proved by direct computation. In particular, it is necessary to use the fact that

$$\begin{aligned} z\cdot e_1 &= z\chi_F^{(n)}(z)\cdot g, \\ &= (\chi_F(z) - \alpha_n)\cdot g, \\ &= -\alpha_n\cdot e_n.] \end{aligned}$$

Note that the last row of F in (5.6) consists of the coefficients of χ_F. By definition, $g = e_n$. Hence g as a column vector in K^n has the representation

$$(5.7) \qquad g = \begin{bmatrix} 0 \\ \vdots \\ 0 \\ 1 \end{bmatrix}.$$

Conversely, suppose (F, g) have the matrix representation (5.6-7) with respect to some basis in K^n. Then (by direct computation) the rank condition (2.8) is satisfied and therefore (F, g) is completely reachable in both the continuous-time and discrete-time cases (Propositions (2.7) and (2.16)).

We have now proved:

(5.8) PROPOSITION. The pair (F, g) is completely reachable if and only if there is a basis relative to which F is given by (5.6) and g by (5.7).

(5.9) COROLLARY. Given an arbitrary n-th degree polynomial $\lambda(z) = z^n + \beta_1 z^{n-1} + \dots + \beta_n$ in $K[z]$, K = arbitrary field. There exists an n-vector ℓ such that $\lambda = \chi_{F-g\ell'}$ if and only if the pair (F, g) is completely reachable.

PROOF. Suppose that (F, g) is completely reachable. With respect to the same basis (5.5) which exhibits the canonical forms (5.6-7), define

$$(5.10) \qquad \ell = \begin{bmatrix} \beta_n - \alpha_n \\ \vdots \\ \beta_1 - \alpha_1 \end{bmatrix}.$$

Then verify by direct computation that $\lambda = \chi_{F-g\ell'}$.

Conversely, suppose that (F, g) is not completely reachable. Then, recalling Proposition (2.12) (which is an algebraic consequence of (2.8) and hence equally valid for both continuous-time and discrete-time), $\dim X_2 > 0$ and so is also $\deg \chi_{F_{22}}$. Since X_1 is an F-invariant subspace of $X = K^n$, the polynomial $\chi_{F_{11}}$ is independent of the choice of basis in K^n and the same is true then also for $\chi_{F_{22}} = \chi_F / \chi_{F_{11}}$. (In particular, $\chi_{F_{22}}$ does not depend on the arbitrary choice of X_2 in satisfying the condition $X = X_1 \oplus X_2$.) In view of (2.12), we have for all n-vectors ℓ,

$$\chi_{F-g\ell'} = \chi_{F_{11}-g_1\ell_1'} \cdot \chi_{F_{22}}, \quad \deg \chi_{F_{22}} > 0.$$

This contradicts the claim that $\lambda = \chi_{F-g\ell'}$ is true for any λ with suitable choice of ℓ. □

In view of the importance of this last result, we shall rephrase it in purely module theoretic terms:

(5.11) THEOREM. Let K be an arbitrary field and X a cyclic $K[z]$-module with generator g and minimal polynomial χ of degree n. There is a bijection between n-th degree polynomials $\lambda(z) = z^n + \beta_1 z^{n-1} + \dots + \beta_n$ in $K[z]$ and K-homomorphisms $\ell: K^n \to K^n: \chi^{(j)} \cdot g \mapsto \ell_j \cdot g$ ($j = 1, \dots, n$ and $\chi^{(j)}$ defined as in (5.5)) such that λ is the minimal polynomial for the new module structure induced on X by the map $z_*: x \mapsto z \cdot x - \ell(x)$.

Note that in (5.11) $\ell(x)$ corresponds to $g\ell' x$ in (5.10).

The map ℓ in (5.11) defines a control law for the system $\Sigma = (F, g, -)$ corresponding to the module X. The passage from z to z_* is the module-theoretic form of the well-known open-loop to closed-loop transformation used in classical linear control theory.

PROOF. Since the vectors $\chi^{(1)} \cdot g, \dots, \chi^{(n)} \cdot g$ form a basis for K^n, ℓ is clearly a well-defined K-homomorphism. We treat ℓ formally as an element of $K[z]$ (that is, an operator on X is a K-vector space), by writing $\ell \cdot x = \ell(\tilde{\xi} \cdot g)$, where $\tilde{\xi}$ represents the equivalence class $[\xi] = \{\xi: \xi \cdot g = x\}$. Unless identically zero, ℓ is never a $K[z]$-homomorphism and therefore ℓ does not commute with nonunits in $K[z]$.

Define $\ell_j = \beta_j - \alpha_j$, $j = 1, \dots, n$. We prove first that this choice of ℓ implies $\lambda^{(j)}(z - \ell) = \chi^{(j)}(z)$ for $j = 1, \dots, n + 1$. Use induction on j. By definition, $\lambda^{(1)}(z - \ell) = \chi^{(1)}(z)$. In the general case,

$$
\begin{aligned}
\lambda^{(j+1)}(z-\ell)\cdot g &= [(z-\ell)\lambda^{(j)}(z-\ell)+\beta_j]\cdot g && \text{(def. of } \lambda^{(j+1)}\text{)},\\
&= [(z-\ell)\chi^{(j)}(z)+\beta_j]\cdot g && \text{(inductive hypothesis)},\\
&= [z\chi^{(j)}(z)+\beta_j-\ell_j]\cdot g && \text{(def. of } \ell\text{)},\\
&= [z\chi^{(j)}(z)+\alpha_j]\cdot g && \text{(def. of } \ell_j\text{)},\\
&= \chi^{(j+1)}(z)\cdot g && \text{(def. of } \chi^{(j+1)}\text{)}.
\end{aligned}
$$

It follows (case $j = n + 1$) that λ annihilates X regarded as a $K[z_*]$-module. On the other hand, the $\lambda^{(1)}(z_*)\cdot g, \ldots, \lambda^{(n)}(z_*)\cdot g$ is a basis for X as a K-vector space since $\chi^{(1)}(z)\cdot g, \ldots, \chi^{(n)}(z)\cdot g$ was such a basis. So X is cyclic with generator g also as a $K[z_*]$-module. Hence by Propositions (5.1-2) the annihilating ideal of g with respect to the $K[z_*]$-module structure cannot be generated by a polynomial of degree less than n, that is, λ is indeed the minimal polynomial with respect to z_*. The correspondence $\lambda \leftrightarrow \ell$ is obviously bijective. □

The proof immediately implies the following

(5.12) COROLLARY. Let $x = \tilde{\xi}\cdot g$ be any element of X viewed as a $K[z]$-module. Then x has the representation $\tilde{\xi}_*\cdot g$ with respect to the $K[z_*]$-module structure on X, where ξ and ξ_* are related as

$$
\begin{aligned}
\xi(z) &= \sum_{j=1}^{n} \xi_j \chi^{(j)}(z)\cdot g\\
\xi_*(z_*) &= \sum_{j=1}^{n} \xi_j \chi_*^{(j)}(z_*)g.
\end{aligned}
$$

So the open-loop/closed-loop transformation is essentially a change in the canonical basis, provided X is cyclic.

It is interesting that the $\chi^{(j)}$ have long been known in Algebra (they are related to the Tschirnhausen transformation discussed extensively by WEBER [1898, §46, 54, 74, 85, 96]), but their present (very natural) use in module theory seems to be new.

**Theorem (5.11) may be viewed as the central special case of Theorem A of the Introduction. Let us restate the latter in precise form as follows:

(5.13) THEOREM. Given an arbitrary n-th degree polynomial $\lambda(z) = z^n + \beta_1 z^{n-1} + \dots + \beta_n$ in $K[z]$, K = arbitrary field. There exists an $n \times m$ matrix L over K such that $\chi_{F-GL'} = \lambda$ if and only if (F, G) is completely reachable.

For some time, this result had the status of a well-known folk theorem, considered to be a straightforward consequence of (5.9). The latter has been discovered independently by many people. (I first heard of it in 1958, proposed as a conjecture by J. E. Bertram and proved soon afterwards by the so-called root-locus method.) Indeed, the passage from (5.11) to (5.13) is primarily a technical problem. A proof of (5.13) was given by LANGENHOP [1964] and subsequently simplified by WONHAM [1967]. The first proof was (unnecessarily) very long, but the second proof is also unsatisfactory; since it depends on arguments using a splitting field of K

**The material between these marks was added after the Summer School.

and fail when K is a finite field. We shall use this situation as an excuse to illustrate the power of the module-theoretic approach and to give a proof of (5.13) valid for arbitrary fields.

The procedure of LANGENHOP and WONHAM rests on the following fact, of which we give a module-theoretic proof:

(5.14) LEMMA. Let K be an arbitrary but infinite field. Let F be cyclic* and (F, G) completely reachable. Then there is an m-vector $a \in K^m$ such that (F, Ga) is also completely reachable.

We begin with a simple remark, which is also useful in reducing the proof of (5.13) to Lemma (5.18).

(5.15) SUBLEMMA. Every submodule of a cyclic module over a principal-ideal domain is cyclic.

PROOF OF (5.14). We use induction on m. The case $m = 1$ is trivial. The general case amounts to the following. Consider the submodule Y of $X = X_F$ generated by the columns $g_1, \ldots, g_{m-1}$ of G. In view of (5.15), Y is cyclic. By the inductive hypothesis, we are given the existence of a cyclic generator of Y of the form $g_y = \alpha_i g_1 + \ldots + \alpha_{m-1} \cdot g_{m-1}$, $\alpha_i \in K$. We must prove: for suitable $\alpha, \beta \in K$ the vector $\alpha \cdot g_Y + \beta \cdot g_m$ is a cyclic generator for X.

----

*Of course, this means that the K[z]-module X_F (see (4.13)) is cyclic.

By hypothesis, X has an (abstract) cyclic generator g_X. By cyclicity we have the representations

$$g_Y = \eta \cdot g_X \quad \text{and} \quad g_m = \mu \cdot g_X, \qquad \eta, \mu \in K[z].$$

Hence our problem is reduced to proving the following: for suitable $\alpha, \beta \in K$ the polynomial $\alpha\eta + \beta\mu$ is a unit in $K[z]/\chi_F K[z]$. This, in turn, is equivalent to proving

$$(5.16) \qquad \alpha\eta + \beta\mu \not\equiv 0 \pmod{\theta_i} \qquad i = 1, \dots, r$$

where $\theta_1, \dots, \theta_r$ in $K[z]$ are the unique prime factors of χ_F. Let $\tilde{}$ mean the representative of least degree of equivalence classes mod θ_i. Then no pair $(\tilde{\eta}_i, \tilde{\mu}_i)$, $i = 1, \dots, r$ can be zero. For if one is, then $\theta_i | (\chi_F, \eta, \mu)$, that is, χ_F/θ_i annihilates the submodule $X' = K[z]g_Y + K[z]g_m$, whence X' is a proper submodule of X, contradicting the fact that (F, G) is completely reachable. If all the $\tilde{\mu}_i$ are zero, then every $\tilde{\eta}_i \neq 0$, so η is a unit in $K[z]/\chi_F K[z]$, and g_Y is already a cyclic generator. So let $\alpha = 1$. Then the condition $\tilde{\eta}_i + \tilde{\beta\mu}_i = 0$ eliminates at most r values of β from consideration. Since K is infinite by hypothesis, there are always some β which satisfy (5.16). □

An essential part of the lemma is the stipulation that $a \in K^m$. The hypothesis "F = cyclic + (F, G) = completely reachable" means that

$$g_X = \alpha_1 g_1 + \dots + \alpha_m g_m, \qquad \alpha_i \in K[z];$$

that is, the lemma is trivially true for some $a \in K^m[z]$ since $g_X = Ga$. But since we want $a \in K$, there must be interaction between vector-space structure and module structure, and for this reason the lemma is nontrivial. As a matter of fact, the lemma is false when K = finite field. The simplest counterexample is provided when (5.12) rules out a <u>single</u> nonzero value of β, thereby ruling out <u>all</u> β.

(5.17) COUNTEREXAMPLE. Let $K = \mathbb{Z}/2\mathbb{Z}$, that is, the ring of integers modulo the prime ideal $2\mathbb{Z}$. Consider

$$F = \begin{bmatrix} 0 & 1 & 0 & 0 & 0 \\ 1 & 1 & 0 & 0 & 0 \\ 0 & 0 & 0 & 1 & 0 \\ 0 & 0 & 0 & 0 & 0 \\ 0 & 0 & 0 & 0 & 1 \end{bmatrix}, \qquad G = \begin{bmatrix} 0 & 0 \\ 1 & 0 \\ 0 & 0 \\ 0 & 1 \\ 1 & 1 \end{bmatrix}.$$

Notice that $X_F = X_1 \oplus X_2 \oplus X_3$ (as a $K[z]$-module), where the minimal polynomials of the direct summands are

$$\begin{aligned} \chi_1(z) &= z^2 + z + 1, \\ \chi_2(z) &= z^2, \\ \chi_3(z) &= z + 1. \end{aligned}$$

All these factors are relatively prime, $(\chi_1, \chi_2, \chi_3) = 1$, hence X is cyclic. Notice also that g_1 generates $X_1 \oplus X_3$ while g_2 generates $X_2 \oplus X_3$. A cyclic generator for X is

$$g_X = \begin{bmatrix} 0 \\ 1 \\ 0 \\ 1 \\ 1 \end{bmatrix}.$$

A simple calculation gives

$$g_1 = z^3 \cdot g_X, \quad g_2 = (z^4 + z^2 + 1) \cdot g_X.$$

Conditions (5.16) are here

$$\alpha \cdot 1 + \beta \cdot 0 \not\equiv 0 \pmod{X_1},$$
$$\alpha \cdot 0 + \beta \cdot 1 \not\equiv 0 \pmod{X_2},$$
$$\alpha \cdot 1 + \beta \cdot 1 \not\equiv 0 \pmod{X_3}.$$

These conditions have no solution in $\mathbb{Z}/2\mathbb{Z}$.

At this point, the following is the situation concerning Theorem (5.13):

(1) Its counterpart, Theorem A of the Introduction, was claimed to be true in the continuous-time case under the hypothesis of complete controllability.

(2) In the discrete-time case (5.13) with the preceding hypothesis Theorem A is false, because of the counterexample: the pair (F = nilpotent, $G = 0$) is completely controllable, but evidently χ_{F-GL}, is independent of L. However, in view of (5.11), Theorem (5.13) might be true also in the discrete-time case if "complete controllability" is replaced by "complete reachability", this modification being immaterial in the continuous-time case.

(3) Because of (5.17), we might expect that a theorem like (5.13) is false for an <u>arbitrary</u> field K.

(4) If our general claim that reachability properties are reflected in module-theoretic properties is true, then (5.13) <u>should hold without assumptions concerning</u> K, because the principal module-theoretic fact, that $K[z]$ = principal ideal domain, is independent of the specific choice of K.

We now proceed to establish Theorem (5.13). That is, special hypotheses on K will turn out to be irrelevant.

PROOF OF (5.13). Necessity is proved exactly as in (5.8). Sufficiency will follow by induction on m, once we have proved it in the special case $m = 2$:

(5.18) LEMMA. *Let* K *be an arbitrary field and let* X *be a* $K[z]$-*module generated by* g_1, g_2. *There is a* K-*homomorphism* ℓ *(of the type defined in* (5.11)) *such that if* $z_* = z - \ell$ *induces a* $K[z_*]$-*module structure on* X *then* X *is cyclic with respect to this structure and is generated by either* $g_1 + g_2$ *or* g_2.

PROOF. Let $Y = K[z]g_1$ and $Z = K[z]g_2$.

Case 1. $Y \cap Z = 0$, that is, $X = Y \oplus Z$. In (5.11) take an ℓ such that $\ell(x) = 0$ for all $x \in Z$. Replacing z by $z_* = z - \ell$ will change the $K[z]$-module structure on Y but preserve that on Z. Further, choose ℓ so that the new minimal polynomial λ on Y is prime to the unchanged minimal polynomial $\chi_{F_Z} = \chi$ on Z. Thus there exist polynomials ν, σ such that $\nu\lambda + \sigma\chi = 1$. By hypothesis, every $x \in X$ has the representation

$$x = y + z = \eta \cdot g_1 + \zeta \cdot g_2.$$

Now verify that

$$\begin{aligned} x &= (\eta\sigma\chi + \zeta\nu\lambda)\cdot(g_1 + g_2), \\ &= \eta\sigma\chi\cdot g_1 + \zeta\nu\lambda\cdot g_2, \\ &= \eta(1 - \nu\lambda)\cdot g_1 + \zeta(1 - \sigma\chi)\cdot g_2, \\ &= \eta\cdot g_1 + \zeta\cdot g_2. \end{aligned}$$

Hence $g_1 + g_2$ is indeed a cyclic generator for X as a $K[z_*]$-module.

Case 2. $Y \cap Z = W \neq 0$. Let $w \in W$. By hypothesis, there is a $\xi \in K[z]$ such that $w = \xi\cdot g_2$ and therefore, by cyclicity of Y, there is also a $\eta \in K[z]$ such that $\xi\cdot g_2 = w = \eta\cdot g_1$. Take same $w \neq 0$. Then if η = unit (mod χ_Y) we are done because $\eta^{-1}\xi\cdot g_2$ generates Y, and so $Z = X$. In the nontrivial case, $\eta \neq$ unit (mod χ_Y). To show: there is a suitable new module structure on X such that $\eta_* =$ unit (mod χ_*), χ_* being the minimal polynomial of X as a $K[z_*]$-module.

The main facts we need are the following:

(5.19) SUBLEMMA. Let χ be a fixed element of $K[z]$ with $\deg \chi = n$, F_χ the companion matrix of χ given by (5.6), X_{F_χ} the cyclic module induced by F_χ, and g a cyclic generator of X_{F_χ}. Then $\eta \in K[z]$ is a unit modulo χ if and only if $\tilde{\eta}\cdot g$ is also a cyclic generator of X_{F_χ}.

PROOF. Obvious. □

(5.20) SUBLEMMA. _Same notations as in_ (5.19). _Write_

$$\tilde{\eta} = \sum_{j=1}^{n} \eta_j \chi^{(j)}(z) \qquad (\chi^{(j)} \text{ defined in } (5.5)).$$

Then $\tilde{\eta}$ _is a unit modulo_ χ _if and only if_

(5.21) $$\det (y, F_\chi y, \dots, F_\chi^{n-1} y) \neq 0,$$

where y _is the column vector_

(5.22) $$y = \begin{bmatrix} \tilde{\eta}_n \\ \cdot \\ \cdot \\ \cdot \\ \tilde{\eta}_1 \end{bmatrix}.$$

PROOF. Since $\chi^{(1)}, \dots, \chi^{(n)}$ is the basis for the K-vector space of all polynomials of degree $< n$, the n-tuple $(\tilde{\eta}_1, \dots, \tilde{\eta}_n)$ is uniquely determined by η. By definition F_χ is the matrix representing the module operator $z\colon x \mapsto z\cdot x$ relative to the special basis $e_1, \dots, e_n$ in X_{F_χ} given by (5.5). Similarly, using one of the module axioms, we verify that

$$\begin{aligned} \tilde{\eta}\cdot g &= \sum_{j=1}^{n} [\tilde{\eta}_j \chi^{(j)}(z)]\cdot g, \\ &= \sum_{j=1}^{n} \tilde{\eta}_j [\chi^{(j)}(z)\cdot g], \\ &\sum_{j=1}^{n} \tilde{\eta}_{n-j+1}\cdot e_j; \end{aligned}$$

in other words, the numerical vector (5.22) represents the abstract vector $\tilde{\eta}\cdot g$ in X_{F_χ} relative to the same basis $e_1, \dots, e_n$. Recall that $\tilde{\eta}\cdot g$ generates X_{F_χ} iff $(F_\chi, \tilde{\eta}(F_\chi)g)$ is complete reachable. By (2.7) the latter condition is equivalent to (5.21). The rest follows from (5.19). □

(5.23) SUBLEMMA. Same notations as in (5.19) and (5.20). Given any nonzero numerical n-vector (5.22), there exists a polynomial χ such that (5.21) is satisfied.

PROOF. Let $\tilde{\eta}_r$ be the first member of the sequence of numbers $\tilde{\eta}_1, \tilde{\eta}_2, \dots$ which is nonzero. Write

$$\chi(z) = z^n + \alpha_1 z^{n-1} + \dots + \alpha_n,$$

and determine the first r coefficients of χ by the rule

$$\begin{bmatrix} \tilde{\eta}_r & \tilde{\eta}_{r+1} & \cdots & \tilde{\eta}_n \\ 0 & \tilde{\eta}_r & \cdots & \tilde{\eta}_{n-1} \\ \cdot & \cdot & & \cdot \\ \cdot & \cdot & & \cdot \\ \cdot & \cdot & & \cdot \\ 0 & 0 & \cdots & \tilde{\eta}_r \end{bmatrix} \begin{bmatrix} \alpha_r \\ \alpha_{r+1} \\ \cdot \\ \cdot \\ \cdot \\ \alpha_n \end{bmatrix} = \begin{bmatrix} 0 \\ 0 \\ \cdot \\ \cdot \\ \cdot \\ 1 \end{bmatrix}.$$

(Since all numbers belong to a field, the required values of $\alpha_r, \dots, \alpha_n$ exist.) Now check, by computation, that these conditions reduce the matrix in (5.21) to the direct sum of two triangular matrices, each with nonzero elements on its diagonal. □

In view of (5.12), it follows from these facts that we can always choose a new $\chi_Y = \chi_\dagger$ such that $\eta_\dagger$ = unit mod $\chi_\dagger$.

The proof of Case 2 is not yet complete, however, because we must still extend the $K[z_*]$-module structure from Y to X. This is easy. Write first $Z = W \oplus Z'$ and then $X = Y \oplus Z'$, where the direct sum is now with respect to the K-module structure of X. Extend ℓ from Y to X by setting $\ell | Z' = 0$. Now we have a new minimal polynomial χ_* defined over X Since $z_* = z_\dagger$ on Y, $\eta_* = \eta_\dagger$. By (5.12), ξ is replaced by some ξ_* such that

$$(5.24) \qquad w = \xi_* \cdot g_1 = \eta_* \cdot g_2,$$

that is, our previous representation of $w \neq 0$ in W induces a similar representation with respect to the new $K[z_*]$-module structure on X. Since η_* is a unit modulo $\chi_\dagger$, we can write

$$\sigma\eta_* = 1 + \tau\chi_\dagger, \quad \text{with } \sigma, \tau \in K[z_*].$$

By (5.24), we have, with respect to the $K[z_*]$-structure,

$$\begin{aligned}(\sigma\xi_*)\cdot g_2 &= \sigma\cdot(\xi_*\cdot g_2),\\ &= \sigma\cdot(\eta_*\cdot g_1),\\ &= (1 + \tau\chi_\dagger)\cdot g_1,\\ &= g_1.\end{aligned}$$

This proves that g_2 generates both Y and Z; that is, g_2 is a cyclic generator for X endowed with the $K[z_*]$-structure. The proof of Lemma (5.18) is now complete. □

It should be clear that Theorem (5.13) is not a purely module-theoretic result, but depends on the interplay between module theory, vector-spaces, and elimination theory (via (5.21)). For instance, the fact that ℓ can be extended from Y to X, which was needed in the proof of Case 2, is a typical vector-space argument.**

There are many open (or forgotten) results concerning cyclic modules which are of interest in system theory. For instance, it is easy to show that an $n \times n$ real matrix is cyclic iff a certain polynomial $\Psi \in \underline{\underline{R}}[z_1, \ldots, z_{n^2}]$ is nonzero at F; the polynomial Ψ is roughly analogous to the polynomial det in the same ring, but, unlike in the latter case, the general form of Ψ does not seem to be known.

We must not terminate this discussion without pointing out another consequence of cyclicity which transcends the module framework. Since X = cyclic with generator g is isomorphic with $K[z]/X_g K[z]$, it is clear that X <u>also has the structure of this commutative ring</u>, that is, the product is defined as

$$x \times y = \xi \cdot g \times \eta \cdot g = (\xi\eta) \cdot g = (\widetilde{\xi\eta}) \cdot g.$$

If X_g = irreducible, then X is even a field. Hence, in particular, X has a galois group. <u>No one has ever given a dynamical interpretation of this galois group</u>. In other words, there are obvious algebraic facts in the theory of dynamical systems which have never been examined from the dynamical point of view. For some related comments in the setting of topological semigroups, see DAY and WALLACE [1967].

6. TRANSFER FUNCTIONS

(6.0) PREAMBLE. There has been a vigorous tradition in engineering (especially in electrical engineering in the United States during 1940-1960) that seeks to phrase all results of the theory of linear constant dynamical systems in the language of the Laplace transform. Textbooks in this area often try to motivate their biased point of view by claiming that "the Laplace transform reduces the analytical problem of solving a differential equation to an algebraic problem". When directed to a mathematician, such claims are highly misleading because the mathematical ideas of the Laplace transform are never in fact used. The ideas which are actually used belong to classical complex function theory: properties of rational functions, the partial-fraction expansion, residue calculus, etc. More importantly, the word "algebraic" is used in engineering in an archaic sense and the actual (modern) algebraic content of engineering education and practice as related to linear systems is very meager. For example, the crucial concept of the transfer function is usually introduced via heuristic arguments based on linearity or "defined" purely formally as "the ratio of Laplace transforms of the output over the input". To do the job right, and to recognize the transfer function as a natural and purely algebraic gadget, requires a drastically new point of view, which is now at hand as the machinery set up in Sections 3-5. The essential idea of our present treatment was first published in KALMAN [1965b].

The first purpose of this section is to give an intrinsically algebraic definition of the transfer function associated with a discrete-time, constant, linear input/output map (see Definition (3.10)). Since the applications of transfer functions are standard, we shall not develop them in detail, but we do want to emphasize their role in relating the classical invariant factor theorem for polynomial matrices to the corresponding module theorem (4.34).

Consider an arbitrary $K[z]$-homomorphism $f: \Omega \to \Gamma$ (see lemma (g) following Theorem (4.2)). Then as a "mathematical object" f is equivalent to the set $\{f(e_j),\ i = 1, \ldots, m,\ e_j$ defined by (4.6)$\}$, since

$$(6.1) \qquad f(\omega) = \sum_{j=1}^{m} \omega_j \cdot f(e_j).$$

(The scalar product on the right is that in the $K[z]$-module Γ, as defined in Section 4.) By definition of Γ, each $f(e_j)$ is a formal power series in z^{-1} with vanishing first term. We shall try to represent these formal power series by ratios of polynomials (which we shall call transfer functions*) and then we can replace formula (6.1) by a certain specially defined product of a ratio of polynomials by a polynomial. Some algebraic sophistication will be needed to find the correct rules of calculations. These "rules" will consititute a rigorous (and simple) version of Heaviside's so-called "calculus". There are no conceptual complications of any sort. (However, we are dodging some difficulties by working solely in discrete-time.)

----

*This entrenched terminology is rather unenlightening in the present algebraic context.

Let $X_f = \Omega/\text{kernel } f$ be the state set of f regarded as a $K[z]$-module. We assume that X_f is a torsion module with nontrivial minimal polynomial ψ. Then, for each $j = 1, \ldots, m$ we have

$$(6.2) \qquad \psi \cdot f(e_j) = f(\psi \cdot e_j) = \eta([\psi \cdot e_j]) = \eta(\psi \cdot [e_j]) = 0.$$

By definition of the module structure on Γ, (6.2) means that the ordinary product of the power series $f(e_j)$ by the polynomial ψ is a (vector) polynomial. Hence (6.2) is equivalent to (notation: no dot = ordinary product)

$$(6.2') \qquad \psi f(e_j) = \theta_j \in K^p[z], \quad j = 1, \ldots, m.$$

Intuitively, we can solve this equation by writing $f(e_j) = \theta_j/\psi$. There are two ways of making this idea rigorous.

Method 1. Define

$$(6.3) \qquad f(e_j) = \theta_j/\psi$$

as the formal division of θ_j by ψ into ascending powers of z^{-1}. Check that the coefficient of z^0 is always 0. Verify by computation that the power series so obtained satisfies (6.2').

Method 2. Multiply both sides of (6.2') by z^{-m}. Write $\hat{\psi}(z^{-1}) = z^{-n}\psi(z)$ and $\hat{\theta}_j(z^{-1}) = z^{-n}\theta(z)$. Then $\hat{\psi} \in K[z^{-1}] \subset K[[z^{-1}]]$ and (6.2') becomes

$$(6.2'') \qquad \hat{\psi} f(e_j) = \hat{\theta}_j \in K^p[z^{-1}].$$

Moreover, the 0-th coefficient of $\hat{\psi}$ is 1 (because of the convention that the leading coefficient of ψ is 1), hence $\hat{\psi}$ is a unit in $K[[z^{-1}]]$ and therefore

$$(6.3') \qquad f(e_j) = \hat{\theta}(z^{-1})\hat{\psi}^{-1}(z^{-1}).$$

Note that (6.3) and (6.3') actually give slightly different definitions of $f(e_j)$, depending on whether we use a transfer function with respect to the variable z or z^{-1}. (Both notations have been used in the engineering literature.) For us the formalism of Method 1 is preferable. (The calculations of Method 1 can be reduced by Method 2 to the better-known calculations of the inverse in the ring $K[[z^{-1}]]$.)

Summarizing, we have the easy but fundamental result:

(6.4) EXISTENCE OF TRANSFER FUNCTIONS. _There is a bijective correspondence between_ $K[z]$-_homomorphisms_ $f: \Omega \to \Gamma$ _with minimal polynomial_ ψ _and transfer function matrices of the type_

$$Z = [\theta_1/\psi, \ldots, \theta_m/\psi],$$

where $\theta_j \in K^p[z]$, $\deg \theta_j < \deg \psi$, _and_ ψ _is the least common denominator of_ Z.

In many contexts, it is preferable to deal with the Z_f corresponding to f rather than with f itself. Because the correspondence is bijective, it is clear that all objects induced by f are well-defined also for Z_f and conversely. Thus, for instance,

$$\dim Z_f \triangleq \dim f \triangleq \dim X_f;$$

$$\psi_Z = \text{least common denominator of } Z,$$

$$= \text{minimal polynomial of } f_Z.$$

(6.5) REMARK. In view of Propositions (4.20-21), the natural realization of Z, namely $X_Z \triangleq X_{f_Z}$, is completely reachable as well as completely observable. Not having this fact available before 1960 has caused a great confusion. Questions such as those resolved by Theorem (5.13) tended to be attacked algorithmically, using special tricks amounting to elementary algebraic manipulations of elements of Z. Very few theoretical results could be conclusively established by this route until the conceptual foundations of the theory of reachability and observability were developed.

The preceding results may be restated as "rules" whereby the values of f may be computed using Z. We have in fact, $f(\omega) = Z\cdot\omega$, where

(6.6) $Z\cdot\omega \triangleq (\widetilde{\psi Z\omega})/\psi$,

= multiply the polynomial matrix ψZ consisting of the numerators of Z with ω, reduce to minimal-degree polynomials modulo ψ and then divide formally by ψ as in Method 1 above.

We can also compute the <u>entire</u> output of the system Σ_Z (that is, all output values following the application of the first nonzero input value) by the rule

(6.7) $Z\omega \triangleq (\psi Z\omega)/\psi$,

= same as above, but do not reduce modulo ψ.

In this second case, the output sequence will begin with a <u>positive</u> power of z. (The coefficients of the positive powers of z are thrown away in the definition of f (see (3.7)) and in the definition of the scalar product in Γ, in order to secure a simple formula for $X_f = \Omega/\text{kernel } f$.)

Many other applications of transfer functions may be found in KALMAN, FALB, and ARBIB [1969, Chapter 10, Section 10].

It is easy to show that the transfer function associated with the system $\Sigma_f = (F, G, H)$ is given by $Z_f = H(zI - F)^{-1}G$. (This is just the formal Laplace transform computed from the constant version of (1.12) by setting $z = d/dt$ or from (1.17) by setting $x(t + 1) = zx(t)$.) Probably the simplest way of computing Z is via the formula

$$6.8) \qquad (zI - F)^{-1} = \sum_{j=0}^{q-1} z^j \psi_F^{(q-i)}(F)(z), \qquad q = \deg \psi,$$

where ψ_F is the minimal polynomial of the matrix F and the superscript denotes the special polynomials defined in (5.5). The matrix identity (6.8) follows at once from the classical scalar identity [WEBER, 1898, §4]

$$\pi(z) - \pi(w) = (z - w) \sum_{j=1}^{q-1} z^j \pi^{(q-i)}(w), \qquad q = \deg \pi,$$

upon setting $w = F$, $\pi = \psi_F$, and invoking the Cayley-Hamilton theorem.

Much of classical linear system theory was concerned with computing Z_f. In the modern context, this problem "factors" into first solving the realization problem $f \to \Sigma_f$ and then applying formula (6.8). See Sections 8 and 9.

One of the mysterious features of Rule (6.6) (as contrasted with the conventional rule (6.7)) is the necessity of reducing modulo ψ. The simplest way of understanding the importance of this aspect of the problem is to show how to relate the module invariant factors occuring in the structure theorem (4.34) to the classical facts concerning the invariant factors of a polynomial matrix.

(6.9) INVARIANT FACTOR THEOREM FOR MATRICES. *Let* P *be a* $p \times m$ *matrix with elements in an arbitrary principal-ideal domain* R. *Then*

(6.10) $$P = A\Pi B,$$

where A *and* B *are* $p \times p$ *and* $m \times m$ *matrices (not necessarily unique) with elements in* R *and* $\det A$, $\det B$ *units in* R, *while*

(6.11) $$\Pi = \operatorname{diag}(\lambda_1, \ldots, \lambda_q, 0, \ldots, 0) \quad \text{with } \lambda_i \in R$$

is unique (up to units in R) *with* $\lambda_i \mid \lambda_{i+1}$, $i = 1, \ldots, q-1$, *and* $q = \operatorname{rank} P$. *The* λ_i *are called the invariant factors of* P.

As anyone would expect, there is a correspondence between the module structure theorem (4.34) and the matrix structure theorem (6.9) and, in particular, between the respective invariant factors $\psi_1, \ldots, \psi_r$ and $\lambda_1, \ldots, \lambda_q$. Let us sketch the standard proof of this fact following CURTIS and REINER [1962, §13.3] who also give a proof of (6.9).

PROOF OF (4.34). Consider the R-homomorphism from R^m onto M given by $\mu: e_i \mapsto g_i$, where the e_i are the standard basis elements of R^m (recall (4.6)) and the g_i generate M. Clearly, $M \approx R^s/N$, where $N = \text{kernel } \mu$. It can be proved that $N \approx R^{\ell}$ is a free submodule of R^m, with a basis of at most $\ell \leqq m$ elements. Write each basis element f_j of N as $\sum p_{ij} \cdot e_i$, $p_{ij} \in R$.

Apply (6.9) to the R-matrix P. Define $\hat{f}_j = \sum c_{ij} \cdot f_i$, $C = B^{-1}$, $\hat{e}_j = \sum a_{ij} \cdot e_i$. By (6.10-11), $\hat{f}_k = \lambda_i \cdot \hat{e}_i$. Hence

$$N = \lambda_1 R \oplus \dots \oplus \lambda_r R.$$

Then, by "direct sum",

$$M \approx R/\lambda_r R \oplus \dots \oplus R/\lambda_1 R \oplus R^{m-\ell}, \quad i = 1, \dots, r.$$

That is, (4.34) holds with $\psi_i = \lambda_i$ and $r = \text{rank } P = \ell$. □

By the same type of calculations, we can prove also

(6.12) THEOREM. Let $\lambda_1, \dots, \lambda_q$ be the invariant factors of ψZ given by (6.9), and let $(\lambda_i, \psi) = \theta_i$, $i = 1, \dots, q$. Then the invariant factors of X_Z are

$$\psi_1 = \psi,$$
$$\psi_2 = \psi/\theta_2,$$
$$\vdots$$
$$\psi_r = \psi/\theta_r,$$

where r is the smallest integer such that $\psi | \lambda_i$ for $i = r + 1, \dots, q = \text{rank } \psi Z$.

PROOF. Consider the $K[z]$-epimorphism $\mu: \Omega \to X_Z: \omega \to [\omega]_Z$. Clearly, $\omega \in [0]_Z = \text{kernel } \mu$ iff $Z \cdot \omega = 0$ (see (6.6)). Equivalently, $(\psi Z)\omega = 0 \pmod{\psi}$. Using the representation whose existence is claimed by (6.9), write $\psi Z = C\Lambda D$ (C, Λ, D = matrices over $K[z]$.) Define $W = D^{-1}\Psi$, where

$$\Psi = \operatorname{diag}(\psi_1, \psi_2, \ldots, \psi_r, 1, \ldots, 1).$$

Then $\Lambda\Psi = 0$, $(\psi Z)W = 0$, and W has clearly maximal rank among $K[z]$-matrices with this property. So the columns of the matrix W constitute a basis for kernel μ. The rest follows easily, as in the proof of (4.34). □

(6.13) REMARK. The preceding proof remains correct, without any modification, if the representation $\psi Z = C\Lambda D$, $\det C$, $\det D$ = units is taken in the ring $K[z]/\psi K[z]$, rather than in $K[z]$. The former representation follows trivially from the latter but may be easier to compute.

(6.14) REMARK. Theorem (6.12) shows how to compute the invariant factors of X_Z from those of ψZ. We must <u>define</u> the invariant factors of Z to be the <u>same</u> as those of X_Z (because of the bijective correspondence $Z \leftrightarrow X_Z$). Consistency with (6.12) demands that we write

$$(6.15) \qquad \lambda_i/\psi = (\lambda_i/\theta_i)/(\psi/\theta_i), \qquad \theta_i = (\lambda_i, \omega),$$

where $/$ is defined as in (6.3). In other words, <u>the ψ_i are the denominators of the scalar transfer function λ_i/ψ after cancellation of all common factors.</u>

Theorems (4.34) and (6.12) do not fully reveal the significance of invariant factors in dynamical systems. Nor is it convenient to deduce all properties of matrix-invariant factors from the representation theorem (6.9). It is interesting that the sharpened results we present below are much in the spirit of the original work of WEIERSTRASS, H. J. S. SMITH, KRONECKER, FROBENIUS, and HENSEL, as summarized in the well-known monograph of MUTH [1899].

(6.16) DEFINITION. *Let* A, B *rectangular matrices over a unique factorization domain* R. $A|B$ *(read:* A *divides* B) *iff there are matrices* V, W *(over* R, *of appropriate sizes) such that* $B = VAW$.

This is of course just the usual definition of "divide" in a ring, specialized to the noncommutative ring of matrices.

The following result [MUTH 1899, Theorems IIIa-b, p. 52] shows that in case of principal-ideal domains the correspondence between matrices and their invariant factors preserves the divide relation (is "functorial" with respect to "divide"):

(6.17) THEOREM. *Let* R *be a principal-ideal domain. Then* $A|B$ *if and only if* $\lambda_i(A)|\lambda_i(B)$ *for all* i.

PROOF. *Sufficiency.* Write the representation (6.10) as

$$A = V_1\Lambda_1W_1, \qquad B = V_2\Lambda_2W_2.$$

By hypothesis, there is a Λ_3 (diagonal) such that $\Lambda_1\Lambda_3 = \Lambda_2$. Hence

$$\begin{aligned} B &= V_2\Lambda_1\Lambda_3W_2, \\ &= V_2V_1^{-1}AV_1\Lambda_1W_1W_1^{-1}\Lambda_3W_2, \\ &= (V_2V_1^{-1})A(W_1^{-1}\Lambda_3W_2). \end{aligned}$$

<u>Necessity</u>. This is just the following

(6.18) LEMMA. <u>For an arbitrary unique-factorization domain</u> R, $A \mid B$ <u>implies</u> $\lambda_i(A) \mid \lambda_i(B)$.

PROOF. By elementary determinant manipulations, as in MUTH [1899, Theorem II, p. 16-17]. □

This completes the proof of Theorem (6.17) □

(6.19) REMARK. Since (6.9) does not apply (why?) to unique factorization domains, for purposes of using Lemma (6.18) we need WEIERSTRASS's definition of invariant factors: if $\Delta_j(A)$ = greatest common factor of all $j \times j$ minors of a matrix A, with $\Delta_0(A) = 1$, then $\lambda_i(A) = \Delta_i(A)/\Delta_{i-1}(A)$. Of course, this definition can be shown to be equivalent (over principal-ideal domains) to that implied by (6.9).

In analogy with Definition (6.16), let us agree (note inversion!) on

(6.20) DEFINITION. <u>Let</u> Z_1, Z_2 <u>be transfer-function matrices</u> $Z_1 \mid Z_2$ (<u>read</u>: Z_1 <u>divides</u> Z_2) <u>iff there are matrices</u> V, W <u>over</u> $K[z]$ <u>such that</u> $Z_1 = VZ_2W$. (Note that $Z_1 \mid Z_2$ implies at once: $\psi_{Z_1} \mid \psi_{Z_2}$.)

(6.21) THEOREM. $Z_1 \mid Z_2$ <u>if and only if</u> $\psi_i(Z_1) \mid \psi_i(Z_2)$ <u>for all</u> i.

PROOF. This is the natural counterpart of Theorem (6.16), and follows from it by a simple calculation using the definition of $\psi_i(Z)$ given by (6.15). □

(6.22) DEFINITION. $\Sigma_1 | \Sigma_2$ (read: Σ_1 can be simulated by Σ_2) iff $X_{\Sigma_1} | X_{\Sigma_2}$, that is, iff X_{Σ_1} is isomorphic to a submodule of X_{Σ_2} [or isomorphic to a quotient module of X_{Σ_2}].

This definition is also functorially related to the definition of "divide" over a principal ideal domain R because of the following standard result:

(6.23) THEOREM. Let R be a principal-ideal domain and X, Y R-modules. Then Y is (isomorphic) to a submodule or quotient module of X if and only if

$$\psi_i(Y) | \psi_i(X), \quad i = 1, \ldots, r(Y) \leqq r(X).$$

PROOF. Sufficiency. Take both X and Y in canonical form (4.34), with $x_1, \ldots, x_{r(X)}$ generating the cyclic pieces of X, and $y_1, \ldots, y_{r(X)}$ (with $y_i = 0$ if $i > r(Y)$) those of Y. The assignment $y_i \mapsto (\psi_i(X)/\psi_i(Y)) \cdot x_i$ defines a monomorphism $Y \to X$, that is, exhibits Y as (isomorphic to) a submodule of X. Similarly, the assignment $x_i \mapsto y_i$ defines an epimorphism $X \to Y$ exhibiting Y as (isomorphic to) a quotient module of X.

Necessity (following BOURBAKI [Algèbre, Chapter 7 (2e ed.), Section 4, Exercise 8]). Let Y be a submodule of X. By (4.34), $X \approx L/N$ where L, N are free R-modules. By a classical isomorphism theorem, Y is isomorphic to a quotient module M/N, where $L \supset M \supset N$ and M is free (since submodules of a free module are free).

From the last relation, $r(Y) \leqq r(X)$. Now observe, again using (4.34) that, for any R-module X and any $\pi \in R$,

$$r(\pi X) < k \implies \psi_k(X) \mid \pi$$

and therefore

$$R\psi_k(X) = \text{ideal generated by } \{\pi : \; r(\pi X) < k\}.$$

Since πY is a submodule of πX for all $\pi \in R$, it follows that $R\psi_k(X) \supset R\psi_k(Y)$, and the proof is complete for the case when Y is a submodule of X. The proof of the other case is similar. □

(6.24) COROLLARY. $\psi_i(Z_\Sigma) \mid \psi_i(\Sigma)$, $i = 1, \ldots, r(Z_\Sigma)$.

PROOF. Immediate from the fact that X_{Z_Σ} is a submodule of Σ (see Section 7). □

Now we can summarize main results of this section as the

(6.25) PRIME DECOMPOSITION THEOREM FOR LINEAR DYNAMICAL SYSTEMS. _The following conditions are equivalent:_

(i) Z_1 _divides_ Z_2.

(ii) $\psi_i(Z_1)$ _divides_ $\psi_i(Z_2)$ _for all_ i.

(iii) Σ_{Z_1} _can be simulated by_ Σ_{Z_2}.

PROOF. This follows by combining Theorem (6.21) with Theorem (6.23), since $\psi_i(Z) = \psi_i(\Sigma_Z)$ by definition. □

(6.26) INTERPRETATION. The definition of $Z_1 | Z_2$ means, in system-theoretic terms, that the inputs and outputs of the machine whose transfer function is Z_2 are to be "recoded": the original input ω_2 is replaced by an input $\omega_2 = B(z)\omega_1$ and the output γ_2 is replaced by an output $\gamma_1 = A(z)\gamma_2$; with these "coding" operations, Σ_2 will act like a machine with transfer function Z_1. In view of the definition of a transfer function, the equation $Z_1 = AZ_2B$ is always satisfied whenever A, B are replaced by $\tilde{A}$, $\tilde{B}$ (reduced modulo ψ_{Z_2}). This means that the coding operations can be carried out physically given a <u>delay</u> of $d = \deg \psi_{Z_2}$ units of time (or more). No feedback is involved in coding, it is merely necessary to store the d last elements of the input and output sequences. Hence, in view of Theorem (6.25) and Corollary (6.24), we can say that <u>it is possible to alter the dynamical behavior of a system Σ_2 arbitrarily by external coding involving delay but not feedback if and only if the invariant factors of the desired external behavior (Z_1) are divisors of invariant factors of the external behavior (Z_{Σ_2}) of the given system</u>. The invariant factors may be called the PRIMES of linear systems: they represent the atoms of system behavior which cannot be simulated from smaller units using arbitrary but feedback-free coding. In fact, there is a close (bot not isomorphic) relationship between the Krohn-Rhodes primes of automata theory (see KALMAN, FALB, and ARBIB [1969, Chapters 7-9]) and ours. A full treatment of this part of linear system theory will be published elsewhere.

7. ABSTRACT THEORY OF REALIZATIONS

The purpose of this short section is to review and expand those portions of the previous discussion which are relevant to the detailed theory of realizations to be presented in Sections 8 and 9. The same issues are examined (from a different point of view) also in KALMAN, FALB, and ARBIB [1969].

Let $f: \Omega \to \Gamma$ be a fixed input/output map. Let us recall the construction of X_f, as a set and as carrying a $K[z]$-module structure (Sections 3 and 4). It is clear that (i) $f = \iota_f \circ \mu_f$, where

$$\mu_f: \Omega \to X_f: \omega \mapsto [\omega],$$
$$\iota_f: X_f \to \Gamma: [\omega]_f \mapsto f(\omega)$$

are $K[z]$-homomorphisms, and (ii) μ_f = epimorphism while ι_f = monomorphism. We have also seen that

$$(7.1)\quad \begin{cases} \mu_f = \text{epimorphism} \iff X_f \text{ is completely reachable;} \\ \iota_f = \text{monomorphism} \iff X_f \text{ is completely observable.} \end{cases}$$

These facts set up a "functor" between system-theoretic notions and algebra which characterize X_f uniquely. Consequently, it is desirable to replace also our system-theoretic definition of a realization (3.12) by a purely algebraic one:

(7.2) DEFINITION. A realization of a $K[z]$-homomorphism $f: \Omega \to \Gamma$ is any factorization f that is, any commutative diagram

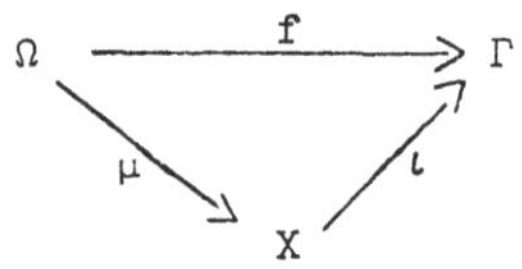

of K[z]-homomorphisms. The K[z]-module X is called the state module of the realization. A realization is canonical iff it is completely reachable and completely observable, that is, μ is surjective and ι is injective.

A realization always exists because we can take $X = \Omega$, $\mu = 1_\Omega$, $\iota = f$ (or $X = \Gamma$, $\mu = f$, $\iota = 1_\Gamma$).

(7.3) REMARK. It is clear that a realization in the sense of (3.12) can always be obtained from a realization given by (7.2). In fact, define $\Sigma = (F, G, H)$ by

$$F\colon X \to X\colon x \mapsto z\cdot x,$$

$$G = \mu \text{ restricted to the submodule } \{\omega\colon |\omega| = 1\}.$$

$$H = \iota \text{ followed by the projection } \gamma \mapsto \gamma(1).$$

It is easily verified that these rules will define a system with f, x = f. Given any such Σ, it is also clear that the rules

$$X = X_\Sigma,$$

$$\mu\colon \omega \mapsto \sum_{t \leqq 0} F_\Sigma^{-t} G_\Sigma \omega(t),$$

$$\nu\colon x \mapsto (H_\Sigma x, H_\Sigma F_\Sigma x, \ldots)$$

define a factorization of f. Hence the correspondence between (3.12) and (7.2) is bijective.

The quickest way to exploit the algebraic consequences of our definition (7.2) is via the following arrow-theoretic fact:

(7.4) ZEIGER FILL-IN LEMMA. *Let* A, B, C, D *be sets and* α, β, γ, *and* δ *set maps for which the following diagram commutes:*

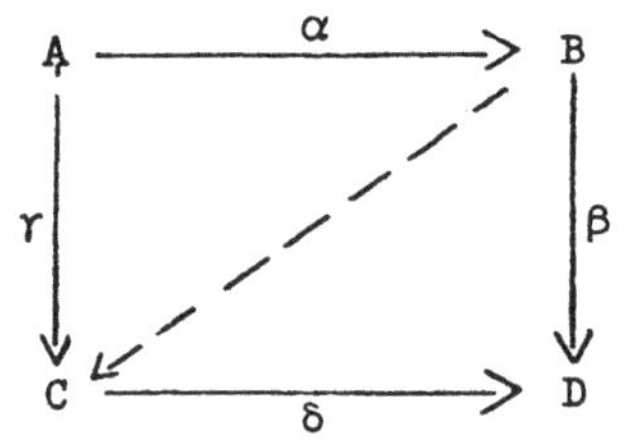

If α *is surjective and* δ *is injective, there exists a unique set map* φ *corresponding to the dashed arrow which preserves commutativity.*

This follows by straightforward "diagram-chasing", which proves at the same time the

(7.5) COROLLARY. *The claim of the lemma remains valid if "sets" are replaced by "R-modules" and "set maps" by "R-homomorphisms".*

Applying the module version of the lemma twice, we get

(7.6) PROPOSITION. *Consider any two canonical realizations of a fixed* f: *the corresponding state-sets are isomorphic as* K[z]-*modules.*

Since every K[z]-module is automatically also a K-vector space, (7.6) shows that the two state sets are K-isomorphic, that is, have the same dimension as vector spaces. The fact that they are also K[z]-isomorphic implies, via Theorem (4.34), that they have the same invariant factors. We have already employed the convention that (in view of the bijection between f and Σ_f), the invariant factors of f and X_f are to be identified. In view of (7.6), this is now a general fact, not dependent on the special construction used to get X_f. We can therefore restate (7.6) as the

(7.7) ISOMORPHISM THEOREM FOR CANONICAL REALIZATIONS. _Any two canonical realizations of a fixed_ f _have isomorphic state modules. The state module of a canonical realization is uniquely characterized (up to isomorphism) by its invariant factors, which may be also viewed as those of_ f.

A simple exercise proves also

(7.8) PROPOSITION. _If_ X _is the state module of a canonical realization_ f, _then_ dim X _(as a vector space) is minimum in the class of all realizations of_ f.

This result has been used in some of the literature to justify the terminology "minimal realization" as equivalent to "canonical realization". We shall see in Section 9 that the two notions are not always equivalent; we prefer to view (7.2) as the basic definition and (7.8) as a derived fact.

(7.9) REMARK. Theorem (7.7) constitutes a proof of the previously claimed (4.24). To be more explicit: if $\Sigma = (F, G, H)$ and $\hat{\Sigma} = (\hat{F}, \hat{G}, \hat{H})$ are two triples of matrices defining canonical realizations of the same f, then (7.7) implies the existence of a vector-space isomorphism $A: X \to \hat{X}$ such that

$$
(7.10) \quad \begin{aligned} \hat{F} &= AFA^{-1}, \\ \hat{G} &= AG, \\ \hat{H} &= HA^{-1}. \end{aligned}
$$

If we identify X and $\hat{X}$ then A is simply a basis change and it follows that _the class of all matrix triples which are canonical realizations of a fixed_ f _is isomorphic with the general linear group over_ X.

The actual computation of a canonical realization, that is, of the abstract Nerode equivalence classes $[\omega]_f$, require a considerable amount of applied-mathematical machinery, which will be developed in the next section. The critical hypthesis is the existence of a factorization of f such that $\dim X < \infty$. (this is sometimes expressed by saying that f has _finite rank_.) Given any such realization, it is possible to obtain a canonical one by a process of reduction. More precisely, we have

(7.11) THEOREM. _Every realization of_ f _with state module_ X _contains a subquotient (a quotient of a submodule, or equivalently, a submodule of a quotient)_ X_* _of_ X _which is the state-module of a canonical realization of_ f.

PROOF. The reachable states X_r = image μ are a submodule of X and so are the unobservable states X_o = kernel ι. Hence $X_* \approx X_r/X_r \cap X_o$ is a subquotient of X. It follows immediately that X_* is a canonical state-module for f. [The proof may be visualized via the following commutative diagram, where the j's and p's are canonical injections and projections.] □

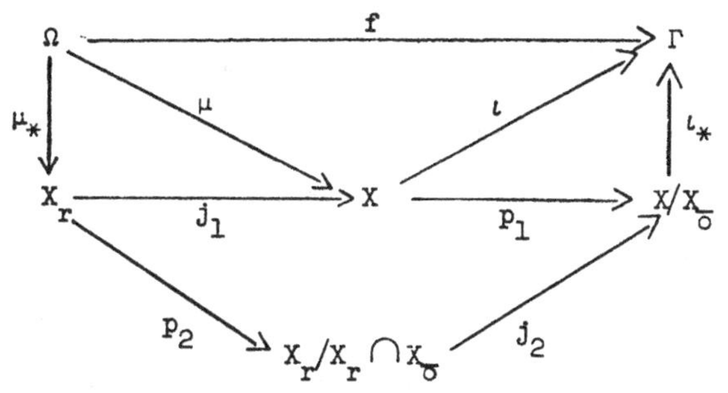

(7.12) REMARK. Since any subquotient of X is isomorphic to a submodule (or a quotient module) of X, it follows from Theorem (6.23) that X can be state-state module of a realization only if $\psi_i(f) | \psi_i(X)$ for all i (recall also Corollary (6.24)). This condition, however, is not enough since the ψ_i are invariants of module isomorphisms and not isomorphisms of the commutative diagram (7.2).

The preceding discussion should be kept in mind to gain an overview of the algorithms to be developed in the next sections.

8. CONSTRUCTION OF REALIZATIONS

Now we shall develop and generalize the basic algorithm, originally due to B. L. Ho (see HO and KALMAN [1966]), for computing a canonical realization $\Sigma = (F, G, H)$ of a given input/output map f. Most of the discussion will be in the language of matrix algebra.

Notations. Here and in Section 9 boldface capital letters* will denote block matrices or sequences of matrices; finite block matrices will be denoted by small Greek subscripts on boldface capitals; the elements of such matrices will be denoted by ordinary capitals. This is intended to make the practical aspects of the computations self-evident; no further explanations will be made.

Let $f: \Omega \to \Gamma$ be a given, fixed $K[z]$-homomorphism. Using only the K-linearity of f we have that

$$(8.1) \qquad f(\omega)(1) = \sum_{t \leqq 0} A_{-t+1}\omega(t),$$

where the A_k $(k > 0)$ are $p \times m$ matrices over the fixed field K. We denote the totality of these matrices by

$$\mathbf{A}(f) = (A_1, A_2, \dots).$$

Then it is clear that the specification of a $K[z]$-homomorphism f is equivalent to the specification of its matrix sequence $\mathbf{A}(f)$. Moreover, if Σ realizes f (8.1) can be written explicitly as

$$(8.2) \qquad f(\omega)(1) = \sum_{t \leqq 0} HF^{-t}G\omega(t).$$

----

*Note to Printer: Indicated by double underline.

Comparing (8.1) and (8.2) we can translate (3.12) into an equivalent matrix-language

(8.3) DEFINITION. A dynamical system $\Sigma = (F, G, H)$ realizes a (matrix) infinite sequence $\underline{\underline{A}}$ iff the relation

$$A_{k+1} = HF^kG, \quad k = 0, 1, 2, \ldots$$

is satisfied.

Let us now try to obtain also a matrix criterion for an infinite sequence $\underline{\underline{A}}$ to have a finite-dimensional realization. The simplest way to do that is to first write down a matrix representation for the map $f: \Omega \to \Gamma$. So let

$$\underline{\underline{H}}(\underline{\underline{A}}) = \begin{bmatrix} A_1 & A_2 & A_3 & \cdots \\ A_2 & A_3 & A_4 & \cdots \\ A_3 & A_4 & A_5 & \cdots \\ \cdot & \cdot & \cdot & \\ \cdot & \cdot & \cdot & \\ \cdot & \cdot & \cdot & \end{bmatrix},$$

and verify that $\underline{\underline{H}}(\underline{\underline{A}}(f))$ represents f when $\omega \in \Omega$ is viewed as an ∞ column vector with elements $(\omega_1(0), \ldots, \omega_m(0), \omega_1(1), \ldots)$. Classically, $\underline{\underline{H}}(\underline{\underline{A}})$ is known as the (infinite) Hankel matrix associated with $\underline{\underline{A}}$. We denote by $\underline{\underline{H}}_{\mu,\nu}$ the $\mu \times \nu$ block submatrix of $\underline{\underline{H}}$ appearing in the upper left-hand corner of $\underline{\underline{H}}$.

(8.4) PROPOSITION. Let Σ be any realization of $\underline{\underline{A}}$. Then

$$\operatorname{rank} \underline{\underline{H}}_{\mu,\nu}(\underline{\underline{A}}) \leqq \dim \Sigma \quad \text{for all} \quad \mu, \nu \geqq 1.$$

(8.5) COROLLARY. An infinite sequence $\underline{\underline{A}}$ has a finite-dimensional realization only if rank $\underline{\underline{H}}_{\mu,\nu}(\underline{\underline{A}})$ is constant for all μ, ν sufficiently large.

PROOF. If $\dim \Sigma = \infty$, the claim of the proposition is vacuous (although formally correct!). Assume therefore that $\dim \Sigma < \infty$ and define from Σ the finite block matrices

$$\underline{\underline{R}}_{\nu} = [G, FG, \ldots, F^{\nu-1}G]$$

and

$$\underline{\underline{O}}_{\mu} = [H', H'F', \ldots, H'(F')^{\mu-1}].$$

Then

$$\underline{\underline{O}}'_{\mu}\underline{\underline{R}}_{\nu} = \underline{\underline{H}}_{\mu,\nu}(\underline{\underline{A}})$$

by the definition (8.3) of a realization. It is clear that rank $\underline{\underline{R}}_{\nu}$ and rank $\underline{\underline{O}}_{\mu}$ are at most $n = \dim \Sigma$. Thus our claim is reduced to the standard matrix fact

$$\operatorname{rank}(AB) \leqq \min\{\operatorname{rank} A, \operatorname{rank} B\}. \qquad \square$$

Our next objective is the proof of the converse of the corollary. This can be done in several ways. The original proof is due to HO and KALMAN [1966]; similar results were obtained independently and concurrently by YOULA and TISSI [1966] as well as by SILVERMAN [1966]. Two different proofs are analyzed and compared in KALMAN, FALB, and ARBIB [1969, Chapter 10, Section 11]. All proofs depend on certain finiteness arguments. We shall give here a variant of the proof developed in HO and KALMAN [1969].

(8.6) DEFINITION. The infinite Hankel matrix $\underline{\underline{H}}$ associated with the sequence $\underline{\underline{A}}$ has finite length $\lambda = (\lambda', \lambda'')$ iff one of the following two equivalent conditions holds:

$$\lambda' = \min \{\ell' : \operatorname{rank} \underline{\underline{H}}_{\ell',\nu} = \operatorname{rank} \underline{\underline{H}}_{\ell'+\kappa,\nu} \text{ for all } \kappa, \nu = 1, 2, \ldots \} < \infty$$

or

$$\lambda'' = \{\min \ell'' : \operatorname{rank} \underline{\underline{H}}_{\mu,\ell''} = \operatorname{rank} \underline{\underline{H}}_{\mu,\ell''+\kappa} \text{ for all } \kappa, \mu = 1, 2, \ldots \} < \infty.$$

λ' is the row length of $\underline{\underline{H}}$ and λ'' is the column length of $\underline{\underline{H}}$.

The equivalence of the two conditions is immediate from the equality of the row rank and column rank of a finite matrix. The proof of the following result (not needed in the sequel) is left for the reader as an exercise in familiarizing himself with the special pattern of the elements of a Hankel matrix:

(8.7) PROPOSITION. For any $\underline{\underline{H}}$, the following inequalities are either both true [$\underline{\underline{H}}$ has finite length] or both false [otherwise]:

$$\lambda' \leqq \operatorname{rank} \underline{\underline{H}}_{m\lambda'',\lambda''} \leqq m\lambda'',$$
$$\lambda'' \leqq \operatorname{rank} \underline{\underline{H}}_{\lambda',p\lambda'} \leqq p\lambda',$$

The most direct consequence of the finiteness condition given by (8.6) is the existence of a finite-dimensional representation $\underline{\underline{S}}$ and $\underline{\underline{Z}}$ of the shift operator σ_A acting on a sequence $\underline{\underline{A}}$. The "operand" will be the Hankel matrix associated with a given $\underline{\underline{A}}$. As we shall see soon, this representation of the shift operator induces a rule for computing the matrix F of a realization of $\underline{\underline{A}}$. This is exactly what we would expect: module theory tells us that, loosely speaking,

$$\underline{\underline{S}},\ \underline{\underline{Z}} \approx \sigma_A \approx z \approx F.$$

(8.7) DEFINITION. <u>The shift operator</u> σ_A <u>on an infinite sequence</u> $\underline{\underline{A}}$ <u>is given by</u>

$$\sigma_A^k\colon (A_1, A_2, \ldots) \mapsto (A_{1+k}, A_{2+k}, \ldots);$$

<u>the corresponding shift operator on Hankel matrices is then</u>

$$\sigma_H^k\colon \underline{\underline{H}}(\underline{\underline{A}}) \mapsto \underline{\underline{H}}(\sigma_A^k \underline{\underline{A}}).$$

(Of course, σ_H is well-defined also on submatrices of a Hankel matrix.)

(8.8) MAIN LEMMA. <u>A Hankel matrix</u> $\underline{\underline{H}}$ <u>associated with an infinite sequence</u> $\underline{\underline{A}}$ <u>has finite length if and only if the shift operator</u> σ_H <u>has finite-dimensional left and right matrix representations</u>. <u>Precisely:</u> $\underline{\underline{H}}$ <u>has finite length</u> $\lambda = (\lambda', \lambda'')$ <u>if and only if there exist</u> $\ell' \times \ell'$ <u>and</u> $\ell'' \times \ell''$ <u>block matrices</u> $\underline{\underline{S}}$ <u>and</u> $\underline{\underline{Z}}$ <u>such that</u>

$$\text{(8.9)}\qquad \begin{aligned} \sigma_H^k \underline{\underline{H}}_{\ell',\ell''}(\underline{\underline{A}}) &= \underline{\underline{S}}\, \underline{\underline{H}}_{\ell',\ell''}(\underline{\underline{A}}), \\ &= \underline{\underline{H}}_{\ell',\ell''}(\underline{\underline{A}})\underline{\underline{Z}}^k, \end{aligned}$$

<u>and furthermore the minimum size of these matrices satisfying</u> (8.9) <u>is</u> $\lambda' \times \lambda'$ <u>and</u> $\lambda'' \times \lambda''$.

PROOF. <u>Sufficiency</u>. Take any $\ell'' \times \ell''$ block matrix $\underline{\underline{Z}}$ which satisfies (8.9). Compute the last column of $\underline{\underline{H}}_{\mu,\ell''}\underline{\underline{Z}}$:

(8.10) $$A_{j+\ell''+1} = A_{j+1}Z_{1\ell''} + A_{j+2}Z_{2\ell''} + \dots + A_{j+\ell''}Z_{\ell''\ell''}$$

for all $j = 0, 1, \dots$ (where $Z_{\mu\nu}$ is the $(\mu, \nu)^{th}$ element block of $\underline{\underline{Z}}$). Relation (8.10) proves that

$$\text{rank } \underline{\underline{H}}_{\kappa+1,\ell''} = \text{rank } \underline{\underline{H}}_{\kappa+1,\ell''} \quad \text{for all } \kappa = 0, 1, \dots ;$$

the general case follows by repetition of the same argument. Hence the existence of the claimed $\underline{\underline{Z}}$ implies that the column length λ'' of $\underline{\underline{H}}$ cannot exceed the size of $\underline{\underline{Z}}$. If actually λ'' is smaller than the size of the smallest $\underline{\underline{Z}}$ which works in (8.9), we get a contradiction from the necessity part of the proof. The claims concerning $\underline{\underline{S}}$ are proved by a strictly dual argument.

<u>Necessity</u>. By the definition of λ'', each column of the $(\lambda'' + 1)^{th}$ block column of $\underline{\underline{H}}_{\mu,\lambda''+1}$ is linearly dependent on the columns of the preceding block columns of $\underline{\underline{H}}_{\mu,\lambda''+1}$; moreover, this property is true for all integers μ, no matter how large. So there exist $m \times m$ matrices $Z_1, \dots, Z_{\lambda''}$ such that the relation

(8.11) $$A_{j+1}Z_{\lambda''} + A_{j+2}Z_{\lambda''+1} + \dots + A_{j+\lambda''}Z_1 = A_{j+1+\lambda''}$$

holds identically for all $j = 0, 1, \dots$. Now define $\underline{\underline{Z}}$ to be an $\lambda'' \times \lambda''$ block companion matrix of $m \times m$ block made up from the Z_i just defined:

$$Z = \begin{bmatrix} 0 & 0 & 0 & \cdots & 0 & Z_{\lambda''} \\ I & 0 & 0 & \cdots & 0 & Z_{\lambda''-1} \\ 0 & I & 0 & \cdots & 0 & Z_{\lambda''-2} \\ \cdot & \cdot & \cdot & & \cdot & \cdot \\ \cdot & \cdot & \cdot & & \cdot & \cdot \\ \cdot & \cdot & \cdot & & \cdot & \cdot \\ 0 & 0 & 0 & \cdots & 0 & Z_2 \\ 0 & 0 & 0 & \cdots & I & Z_1 \end{bmatrix}.$$

The verification of (8.9) is immediate, using (8.11). The existence of $\lambda' \times \lambda'$ block matrix $\underline{\underline{S}}$ verifying (8.9) follows by a strictly dual argument. □

Now we have enough material on hand to prove the strong version of Corollary (8.5):

(8.12) THEOREM. <u>An infinite sequence</u> $\underline{\underline{A}}$ <u>has a finite-dimensional realization of dimension</u> n <u>if and only if the associated Hankel matrix</u> $\underline{\underline{H}}$ <u>has finite length</u> $\lambda = (\lambda', \lambda'')$.

PROOF. <u>Sufficiency</u>. Let $\underline{\underline{E}}_{\lambda'',1}$ be a $\lambda'' \times 1$ block column matrix whose first block element is an $m \times m$ unit matrix and the other blocks are $m \times m$ zero matrices. Using (8.9) with $\ell'' = \lambda''$, define

$$(8.13) \qquad \Sigma = \begin{cases} F = \underline{\underline{Z}}, \\ G = \underline{\underline{E}}_{\lambda'',1}, \\ H = \underline{\underline{H}}_{1,\lambda''}. \end{cases}$$

Then, for all $k \geq 0$, compute

$$\begin{aligned} HF^kG &= \underline{\underline{H}}_{1,\lambda''}\underline{\underline{Z}}^k\underline{\underline{E}}_{\lambda'',1} \\ &= \sigma^k_{H}\underline{\underline{H}}_{1,\lambda''}\underline{\underline{E}}_{\lambda'',1}; \end{aligned}$$

the second step uses (8.9). By definition of σ_A and $\underline{\underline{E}}$, the last matrix is just the $(1, 1)^{th}$ element of $\underline{\underline{H}}(\sigma^k_A(\underline{\underline{A}}))$, namely A_{1+k}. Hence the given Σ is a realization of $\underline{\underline{A}}$.

Necessity. This is immediate from Corollary (8.5). □

Now we want to attack the problem of finding a canonical realization of $\underline{\underline{A}}$, since the realization given by (8.13) is usually very far from canonical. Our succeeding considerations here and in Section 9 are made more transparent if we digress for a moment to establish another consequence of (8.8).

By outrageous abuse of language, we shall say that $\underline{\underline{A}}$ has finite length iff $\underline{\underline{H}}(\underline{\underline{A}})$ has finite length. We note

(8.14) DEFINITION. An infinite sequence $\underline{\underline{B}}$ is an extension of order N of (the initial part of) an infinite sequence $\underline{\underline{A}}$ iff $A_k = B_k$ for $k = 1, \ldots, N$.

(8.15) THEOREM. No infinite sequence of finite length (λ', λ'') has distinct length-preserving extensions of any order $N \geq \lambda' + \lambda''$.

PROOF. Suppose $\underline{\underline{B}}$ is a length-preserving extension of order N of $\underline{\underline{A}}$, the length of both sequences being (λ', λ''), with $N \geq \lambda' + \lambda''$. By (8.8), both sequences satisfy relation (8.9), with suitable $\underline{\underline{S}}_A$ and $\underline{\underline{Z}}_B$.

The sequence $\underline{\underline{A}}$ is uniquely determined by $\underline{\underline{S}}_A$ acting on $\underline{\underline{H}}_{\lambda',\lambda''}(\underline{\underline{A}})$ from the left and the sequence $\underline{\underline{B}}$ is uniquely determined by $\underline{\underline{Z}}_B$ acting on the matrix $\underline{\underline{H}}_{\lambda',\lambda''}(\underline{\underline{B}})$ from the right. The two matrices are equal by hypothesis on N. Moreover,

$$\underline{\underline{S}}_A\underline{\underline{H}}_{\lambda',\lambda''}(\underline{\underline{A}}) = \sigma_A\underline{\underline{H}}_{\lambda',\lambda''}(\underline{\underline{A}})$$

and

$$\underline{\underline{H}}_{\lambda',\lambda''}(\underline{\underline{B}})\underline{\underline{Z}}_B = \sigma_B\underline{\underline{H}}_{\lambda',\lambda''}(\underline{\underline{B}})$$

are <u>also</u> equal, since the matrices on the right-hand side depend only on the 2nd, ..., N-th member of each sequence. Using only this fact and the associativity of the matrix product

$$\begin{aligned}
\underline{\underline{H}}_{\lambda',\lambda''}\underline{\underline{Z}}_B^k &= \underline{\underline{H}}_{\lambda',\lambda''}\underline{\underline{Z}}_B\underline{\underline{Z}}_B^{k-1}, \\
&= \underline{\underline{S}}_A\underline{\underline{H}}_{\lambda',\lambda''}\underline{\underline{Z}}_B^{k-1}, \\
&\;\;\vdots \\
&= \underline{\underline{S}}_A^{k-1}\underline{\underline{H}}_{\lambda',\lambda''}\underline{\underline{Z}}_B, \\
&= \underline{\underline{S}}_A^k\underline{\underline{H}}_{\lambda',\lambda''}.
\end{aligned}$$

So $\underline{\underline{A}} = \underline{\underline{B}}$. □

Now we can hope for a realization algorithm which uses only the first $\lambda' + \lambda''$ terms of a sequence of finite length. In fact, we have

(8.16) B. L. HO's REALIZATION ALGORITHM. <u>Consider any infinite sequence</u> $\underline{\underline{A}}$ <u>of finite length with associated Hankel matrix</u> $\underline{\underline{H}}$. <u>The following steps will lead to a canonical realization of</u> $\underline{\underline{A}}$:

(i) *Determine* λ', λ''.

(ii) *Compute* $n = \text{rank}\ \underline{\underline{H}}_{\lambda', \lambda''}$; *in doing so, determine nonsingular* $p\lambda' \times p\lambda'$ *and* $m\lambda'' \times m\lambda''$ *matrices* P, Q *such that*

(8.17)
$$P\underline{\underline{H}}_{\lambda', \lambda''}Q = \begin{bmatrix} I_n & 0 \\ 0 & 0 \end{bmatrix}.$$

(iii) *Compute*

(8.18)
$$\begin{cases} F = R_n P[\sigma_{\underline{\underline{H}}} \underline{\underline{H}}_{\lambda', \lambda''}]QC^n, \\ G = R_n P\underline{\underline{H}}_{\lambda', \lambda''}C^m, \\ H = R_p \underline{\underline{H}}_{\lambda', \lambda''}QC^n, \end{cases}$$

where R_p, C^m *are idempotent "editing" matrices corresponding to the operations "retain only the first* p *rows" and "retain only the first* m *columns".*

We claim the

(8.19) REALIZATION THEOREM FOR INFINITE SEQUENCES. *For any infinite sequence* $\underline{\underline{A}}$ *whose associated Hankel matrix* $\underline{\underline{H}}$ *has finite length* (λ', λ''), *B. L. Ho's formulas* (8.17-18) *yield a canonical realization.*

PROOF. If Σ defined by (8.17-18) is a realization of $\underline{\underline{A}}$, then it is certainly canonical: by (8.4) Σ has minimal dimension in the class of all realizations of $\underline{\underline{A}}$ and so it is canonical by (7.8).

The required verification is interesting. First, drop all subscripts. Observe that $\underline{\underline{H}}^{\#} = QCRP$ is a *pseudo-inverse* of $\underline{\underline{H}}$, that is, $\underline{\underline{H}}\,\underline{\underline{H}}^{\#}\underline{\underline{H}} = \underline{\underline{H}}$. Then, by definition of F, G, H, and $\underline{\underline{H}}^{\#}$,

$$\begin{aligned}
HF^kG &= (R\underline{\underline{H}}QC)(RP[\sigma_{\underline{\underline{H}}}\underline{\underline{H}}]QC)^k(RP\underline{\underline{H}}C),\\
&= R\underline{\underline{H}}(\underline{\underline{H}}^{\#}[\sigma_{\underline{\underline{H}}}\underline{\underline{H}}])^k\underline{\underline{H}}^{\#}\underline{\underline{H}}C;
\end{aligned}$$

by repeated application of (8.9),

$$\begin{aligned}
&= R\underline{\underline{H}}(\underline{\underline{H}}^{\#}\underline{\underline{H}}\,\underline{\underline{Z}})^k\underline{\underline{H}}^{\#}\underline{\underline{H}}C\\
&= R\underline{\underline{S}}\,\underline{\underline{H}}(\underline{\underline{H}}^{\#}\underline{\underline{H}}\,\underline{\underline{Z}})^{k-1}H^{\#}HC,\\
&\quad\vdots\\
&= R\underline{\underline{S}}^k\underline{\underline{H}}\,\underline{\underline{H}}^{\#}\underline{\underline{H}}C,\\
&= R\underline{\underline{S}}^k\underline{\underline{H}}C,\\
&= R[\sigma^k_{\underline{\underline{H}}}\underline{\underline{H}}]C.
\end{aligned}$$

The last equation calls for picking out the first p rows and the first m columns of $\sigma^k_{\underline{\underline{H}}}\underline{\underline{H}}$, which is just A_{1+k}, as required. □

(8.20) COMMENT. This is a considerably sharper result than Theorem (8.12), in two respects:

(i) It is no longer necessary to compute $\underline{\underline{Z}}$: we simply use the matrix $\underline{\underline{H}}_{\lambda',\lambda''}(\sigma_{\underline{\underline{A}}}\underline{\underline{A}})$, which is part of the data of the problem.

(ii) Formulas (8.18) give the desired realization in minimal form: there is no need to reduce (8.13) to a minimal realization (recall here (7.11)).

Notice also that the proof of (8.19) does not require (8.12) but depends (just like the latter) on direct use of (8.8).

An apparently serious limitation of the algorithm (8.16) is the necessity to verify abstractly that "$\underline{\underline{A}}$ has finite length". Of course, this can be done only on the basis of certain special hypotheses on $\underline{\underline{A}}$, given in advance. (Examples: (i) $A_k = 0$ for all $k > q$; (ii) A_k = coefficients of the Taylor expansion of a rational function.) Fortunately, the difficulty is only apparent, for the preceding developments can be sharpened further:

(8.21) FUNDAMENTAL THEOREM OF LINEAR REALIZATION THEORY. Consider any infinite sequence $\underline{\underline{A}}$ and the corresponding Hankel matrix $\underline{\underline{H}}$. Suppose there exist integers ℓ', ℓ'' such that

(8.22) $$\operatorname{rank} \underline{\underline{H}}_{\ell',\ell''}(\underline{\underline{A}}) = \operatorname{rank} \underline{\underline{H}}_{\ell'+1,\ell''}(\underline{\underline{A}}), = \operatorname{rank} \underline{\underline{H}}_{\ell',\ell''+1}(\underline{\underline{A}}).$$

Then there exists unique extension $\hat{\underline{\underline{A}}}$ of $\underline{\underline{A}}$ of order $\ell' + \ell''$ such that $\lambda'_{\hat{\underline{\underline{A}}}} \leq \ell'$ and $\lambda''_{\hat{\underline{\underline{A}}}} \leq \ell''$; moreover, applying formulas (8.17-18) with $\lambda' = \ell'$, $\lambda'' = \ell''$ gives a canonical realization of $\hat{\underline{\underline{A}}}$.

PROOF. Exactly as in the necessity part of the proof of (8.8), condition (8.22) implies the existence of $\underline{\underline{S}}$ and $\underline{\underline{Z}}$ such that

(8.23) $$\sigma_{\underline{\underline{H}}} \underline{\underline{H}}_{\ell',\ell''}(\underline{\underline{A}}) = \underline{\underline{S}}\underline{\underline{H}}_{\ell',\ell''}(\underline{\underline{A}}) = \underline{\underline{H}}_{\ell',\ell''}(\underline{\underline{A}})\underline{\underline{Z}}.$$

Define an extension $\hat{\underline{\underline{A}}}$ of $\underline{\underline{A}}$ of order $\ell' + \ell''$ by

$$\sigma^k_{\underline{\underline{H}}} \underline{\underline{H}}_{\ell',\ell''}(\hat{\underline{\underline{A}}}) \triangleq \underline{\underline{S}}^k \underline{\underline{H}}_{\ell',\ell''}(\underline{\underline{A}}), \quad k > 1.$$

By repeated application of (8.23), it follows that we have also

$$\sigma^k_{\underline{\underline{H}}} \underline{\underline{H}}_{\ell',\ell''}(\hat{\underline{\underline{A}}}) = \underline{\underline{H}}_{\ell',\ell''}(\underline{\underline{A}})\underline{\underline{Z}}^k, \qquad k \geqq 0.$$

Now it is clear, from (8.8), that $\lambda'_{\hat{\underline{\underline{A}}}} \leqq \ell'$ and $\lambda''_{\hat{\underline{\underline{A}}}} \leqq \ell''$. The uniqueness of the extension follows immediately from (8.15). Moreover, Theorem (8.19) is still valid, even though (ℓ', ℓ'') is not necessarily minimal, because the proof of (8.19) depended only on (8.9) and not on the minimality of (ℓ', ℓ''). □

Theorem (8.21) says, in effect, that a canonical realization of some extension of $\underline{\underline{A}}$ is always possible as soon as (8.22) is satisfied. Moreover, (8.22) can be used as a practical criterion for constructing by trial and error a canonical realization of any $\underline{\underline{A}}$ known to have finite length (but without being given λ', λ'').

(8.24) EXAMPLES. (i) There is no scalar infinite sequence $(p = m = 1)$ $\underline{\underline{A}}$ for which (8.22) is never satisfied.

(ii) If $\underline{\underline{H}}_{\ell',\ell''}$ is square and has full rank (for instance, in the scalar case), then (8.22) is automatically satisfied.

(iii) If the algorithm (8.16) is applied without any information concerning condition (8.22), the system Σ defined by (8.18) will always realize some extension of $\underline{\underline{A}}$, at least of order 1. It is not known, however, how to get a simple formula which would determine the maximal order of this extension of $\underline{\underline{A}}$.

The remaining interesting question is then: What can be said if (8.22) is not satisfied for a finite amount of data $A_1, \ldots, A_N$ and any ℓ', ℓ'' satisfying $\ell' + \ell'' = N$. This problem is the topic of the next section.

(8.25) FINAL COMMENT. An essential feature of B. L. Ho's algorithm is that is preserves the block structure of the data of the problem. Of course, one can obtain parallel results by treating $\underline{\underline{H}}_{\ell',\ell''}$ as an ordinary matrix, disregarding its block-Hankel structure. Such a procedure requires looking at a minor of $\underline{\underline{H}}$ of maximum rank, and was described explicitly by SILVERMAN [1966] and SILVERMAN and MEADOWS [1969]. There does not seem to be any obvious computational advantage associated with the second method.

9. THEORY OF PARTIAL REALIZATIONS

In one obvious respect the theory of realizations developed in the previous section is rather unsatisfactory: it is concerned with infinite sequences. From here on we call a system satisfying (8.3) a complete realization, to distinguish it from the practically more interesting case given by

(9.1) DEFINITION. *Let* $\underline{\underline{A}} = (A_1, A_2, \ldots)$ *be an infinite sequence of* $p \times m$ *matrices over a fixed field* K. *A dynamical system* $\Sigma = (F, G, H)$ *is a partial realization of order* r *of* $\underline{\underline{A}}$ *iff*

$$A_{k+1} = HF^kG \quad \text{for} \quad k = 0, 1, \ldots, r.$$

We shall use the same terminology if, instead of an infinite sequence $\underline{\underline{A}}$, we are given merely a finite sequence $\underline{\underline{A}}_s = (A_1, \ldots, A_s)$, $s \geqq r$. The reason for this convention will be clear from the discussion to follow. We shall call the first r terms of $\underline{\underline{A}}$ a partial sequence (of order r).

The concepts of canonical partial realization and minimal partial realization will be understood in exactly the same sense as for a complete realization. We warn the reader, however, that now these two notions will turn out to be inequivalent, in that

$$\text{minimal partial} \Rightarrow \text{canonical partial}$$

but not conversely.

Our main interest will be to determine all equivalence classes of minimal partial realizations; in general, a given sequence will have infinitely many inequivalent minimal partial realizations if r is sufficiently small.

According to the Main Theorem (8.21) of the theory of realizations, the minimal partial realization problem has a unique solution whenever the rank condition (8.22) is satisfied. If the length r of the partial sequence is prescribed a priori, it may well happen that (8.22) does not hold. What to do? Clearly, if we have a minimal partial realization (F, G, H) of order r we can extend the partial sequence of $\underline{\underline{A}}_r$ on which this realization is based to an infinite sequence canonically realized by (F, G, H) simply by setting

$$A_k \triangleq HF^{k-1}G, \quad k > r.$$

Consequently, we have the preliminary

(9.2) PROPOSITION. <u>The determination of a minimal partial realization for</u> $\underline{\underline{A}}_r$ <u>is equivalent to the determination of all extensions of a partial sequence</u> $\underline{\underline{A}}_r$ <u>such that the extended sequence is</u>

(i) <u>finite-dimensional and, more strongly,</u>

(ii) <u>its dimension is minimal in the class of all extensions</u>.

It is trivial to prove that finite-dimensional extensions exist for any partial sequence (of finite length). Hence the problem is immediately reduced to determining extensions which have minimal dimension. The solution of this latter problem consists of two steps. First, we show by a trivial argument that the minimal dimension can be bounded from below by an examination of the Hankel array defined by the partial sequence. Second, and this is rather surprising, we show that the lower bound can be actually attained. For further details, especially the characterization of equivalence classes of the minimal partial realizations, see KALMAN [1969c and 1970b].

(9.3) DEFINITION. *By the Hankel array* $H(A_r)$ *of a partial sequence* A_r *we mean that* $r \times r$ *block Hankel matrix whose* $(i, j)^{th}$ *block is* A_{i+j-1} *if* $i + j - 1 \leq r$ *and undefined otherwise.*

In other words, the Hankel array of a partial sequence A_r consists of block rows and columns made up of subsequences $A_p, \ldots, A_r$ $(1 \leq p \leq r)$ of A_r and blank spaces.

(9.4) PROPOSITION. *Let* $n_o(A_r)$ *be the number of rows of the Hankel array of* A_r *which are linearly independent of the rows above them. Then the dimension of a realization of* A_r *is at least* $n_o(A_r)$.

PROOF. The rank of *any* Hankel matrix of an infinite sequence A is a lower bound on the dimension of *any* realization of A, by Proposition (8.4). By Proposition (9.2), it suffices to consider a suitable extension A of A_r. This implies "filling in" the blank spaces in the Hankel array of A_r. Regardless of how $H(A_r)$ is filled in, the rank of the resulting $r \times r$ block Hankel matrix is bounded from below by $n_o(A_r)$. □

By the block symmetry of the Hankel matrix, we would expect to be able to determine $n_o(A_r)$ by an analogous examination of the columns of the Hankel array of $\underline{\underline{A}}_r$, thereby obtaining the <u>same</u> lower bound. This is indeed true. We prefer not to give a direct proof, since the result will follow as a corollary of the Main Theorem (9.7).

The critical fact is given by the

(9.5) MAIN LEMMA. *For a partial sequence* $\underline{\underline{A}}_r$ *define:*

$\lambda'(\underline{\underline{A}}_r)$ = *smallest integer such that for* $k' > \lambda'$ *every row of* $\underline{\underline{H}}(\underline{\underline{A}}_r)$ *is linearly dependent on the rows above it.*

$\lambda''(\underline{\underline{A}}_r)$ = *smallest integer such that for* $k'' > \lambda''$ *every column in the* k-th *block column of* $\underline{\underline{H}}(\underline{\underline{A}}_r)$ *is linearly dependent on the columns to the left of it.*

Every partial sequence $\underline{\underline{A}}_r$ *may be extended to an infinite sequence* $\underline{\underline{A}}$ *in at least one way such that the condition*

(9.6) $$\operatorname{rank} \underline{\underline{H}}_{\mu,\nu}(\underline{\underline{A}}) = n_o(\underline{\underline{A}}_r) \quad \text{for all} \quad \mu > \lambda'(\underline{\underline{A}}_r),\ \nu > \lambda''(\underline{\underline{A}}_r)$$

is satisfied.

PROOF. The existence of the numbers λ'. λ'' is trivial.

It suffices to show, for arbitrary r, how to select A_{r+1} in such a way that the numbers λ', λ'', and n_o remain constant.

Consider the first row of A_{r+1} and examine in turn all the first rows of the first, second, third, ..., λ'-th block rows in $\underline{\underline{H}}(\underline{\underline{A}}_r)$. If the first row of the first block row is linearly dependent on the rows above it (that is, 0), we fill in the first row of A_{r+1} using this linear dependence (that is, we make the first row of A_{r+1} all zeros). This choice of the first row of A_{r+1} will preserve linear dependencies for the first row of every block row below the second block row, by the definition of the Hankel pattern. If the first row in the first block row is linearly independent of those above (that is, contributes 1 to $n_o(\underline{\underline{A}}_r)$), we pass to the second block row and repeat the procedure. Eventually the first row of some block row will become linearly dependent on those above it, except when $\lambda' = r$; in that case, choose the first row of A_{r+1} to be linearly dependent of the first rows of $A_1, \ldots, A_r$. Repeating this process for the second, third, ... rows of each block row*, eventually A_{r+1} is determined without increasing λ' or n_o.

To complete the proof, we must show that the above definition of A_{r+1} also preserves the value of λ''. That is, we must show that no new independent columns are produced in the Hankel array of $\underline{\underline{A}}_r$ when A_{r+1} is filled in. This is verified immediately by noting that the definition of A_{r+1} implies the conditions

$$\operatorname{rank} \underline{\underline{H}}_{r,1} = \operatorname{rank} \underline{\underline{H}}_{r+1,1},$$

$$\operatorname{rank} \underline{\underline{H}}_{r-1,2} = \operatorname{rank} \underline{\underline{H}}_{r,2},$$

$$\cdots$$

$$\operatorname{rank} \underline{\underline{H}}_{1,r} = \operatorname{rank} \underline{\underline{H}}_{2,r} = \operatorname{rank} \underline{\underline{H}}_{1,r+1}. \qquad \square$$

----

*Of course, now linear dependence in the first step does <u>not</u> imply that the corresponding row of A_{r+1} will be all zeros.

With the aid of this simple but subtle observation, the problem is reduced to that covered by the Main Theorem (8.21) of Section 8. We have:

(9.7) MAIN THEOREM FOR MINIMAL PARTIAL REALIZATIONS.* Let $\underline{\underline{A}}_r$ be a partial sequence. Then:

(i) Every minimal realization of $\underline{\underline{A}}_r$ has dimension $n_o(\underline{\underline{A}}_r)$.

(ii) All minimal realizations may be determined with the aid of B. L. Ho's formulas (8.17-18) with $\lambda' = \lambda'(\underline{\underline{A}}_r)$ and $\lambda'' = \lambda''(\underline{\underline{A}}_r)$ as given by Lemma (9.5).

(iii) If $r \geqq \lambda'(\underline{\underline{A}}_r) + \lambda''(\underline{\underline{A}}_r)$ then the minimal realization is unique. Otherwise there are as many minimal realizations as there are extensions of $\underline{\underline{A}}_r$ satisfying (9.6).

PROOF. By the Main Lemma (9.5), every partial sequence $\underline{\underline{A}}_r$ has at least one infinite extension which preserves λ', λ'' and n_o. So we can apply the (8.21) of the preceding section. It follows that the minimal partial realization is unique if $r \geqq \lambda'(\underline{\underline{A}}_r) + \lambda''(\underline{\underline{A}}_r)$ (the $\lambda'(\underline{\underline{A}}_r) + \lambda''(\underline{\underline{A}}_r) + 1$ Hankel matrix can be filled in completely with the available data); in the contrary case, the minimal extensions will depend on the manner in which the matrices $A_{r+1}, \ldots, A_{\lambda'+\lambda''}$ have been determined (subject to the requirement (9.6)). □

In view of the theorem, we are justified in calling the integer $n_o(\underline{\underline{A}}_r)$ the dimension of $\underline{\underline{A}}_r$.

---

*A similar result was obtained simultaneously and independently by T. Tether (Stanford dissertation, 1969).

(9.8) REMARK. The essential point is that the quantities n_o, λ', and λ'' are uniquely determined already from partial data, irrespective of the possible nonuniqueness of the minimal extensions of the partial sequence. We warn, however, that this result does not generalize to all invariants of the minimal realization. For instance, one cannot determine from $\underline{\underline{A}}_r$ how many cyclic pieces a minimal realization of $\underline{\underline{A}}_r$ will have: some minimal realizations may be cyclic and others may not [KALMAN 1970b].

Finally, let us note also a second consequence of the Main Theorem:

(9.9) COROLLARY. Suppose $n_1(\underline{\underline{A}}_r)$ is the number of independent columns of the Hankel array of $\underline{\underline{A}}_r$ (defined analogously with $n_o(\underline{\underline{A}}_r)$). Then $\dim \underline{\underline{A}}_r = n_1(\underline{\underline{A}}_r)$.

PROOF. If $n_1(\underline{\underline{A}}_r) > n_o(\underline{\underline{A}}_r)$ then, using the Main Theorem, we get a contradiction to the fact that the rank of any Hankel matrix of an infinite sequence is lower bound for the dimension of any realization (Proposition (8.4)). If $n_1(\underline{\underline{A}}_r) < n_o(\underline{\underline{A}}_r)$ then extending $\underline{\underline{A}}_r$ to any $\underline{\underline{A}}_{\lambda'+\lambda''-1}$ we contradict the fact that rank $\underline{\underline{H}}_{\lambda',\lambda''}$ is at least equal to $n_o(\underline{\underline{A}}_r)$. □

In other words, the characteristic property of rank, that counting rank by row or column dependence yields identical results, is preserved even for incomplete Hankel arrays.

It is useful to check a simple case which illustrates some of the technicalities of the proof of the Main Lemma.

(9.10) EXAMPLE. The dimension of $(0, 0, \ldots, 0, A_r)$ is precisely $r \times \rho$, where $\rho = \operatorname{rank} A_r$ and $\lambda' = \lambda'' = r$.

10. GENERAL THEORY OF OBSERVABILITY

In this concluding section, we wish to discuss the problem of observability in a rather general setting: we will not assume linearity, at least in the beginning. This is an ambitious program and leads to many more problems than results. Still, I think it is interesting to give some indication of the difficulties which are conceptual as well as mathematical. This discussion can also serve as an introduction to very recent research [KALMAN 1969a, 1970a] on the observability problem in certain classes of nonlinear systems.

The motivation for this section, as indeed for the whole theory of observability, stems from the writer's discovery [KALMAN 1960a] that the problem of (linear) statistical prediction and filtering can be formulated and resolved very effectively by consistent use of dynamical concepts and methods, and that this whole theory is a strict dual of the theory of optimal control of linear systems with quadratic Lagrangian. For those who are familiar with the standard classical theory of statistical filtering (see, for instance, YAGLOM [1962]), we can summarize the situation very simply by saying that

  Wiener-Kolmogorov filter
+ theory of finite-dimensional linear dynamical systems
= Kalman filter.

For the latter, the original papers are [KALMAN 1960a, 1963a] and [KALMAN and BUCY 1961].

The reader interested in further details and a modern exposition is referred especially to the monograph of KALMAN [1969b].

We shall examine here only one aspect of this theory (which does not involve any stochastic elements): the strict formulation of the "duality principle" between reachability and observability. This principle was formally stated for the first time by KALMAN [1960c], but the pertinent discussion in this paper is limited to the linear case and is somewhat ad-hoc. Aided by research progress since 1960, it is now possible to develop a completely general approach to the "duality principle". We shall do this and, as a by-product, we shall obtain a new and strictly deductive proof of the principle in the now classical linear case.

We shall introduce a general notion of the "dual" system, and use it to replace the problem of observability by an equivalent problem of reachability. In keeping with the point of view of the earlier lectures, we shall view a system in terms of its input/output map f and dualize f (rather than Σ). The constructibility problem will not be of direct interest, since its theory is similar to that of the observability problem.

Let Ω, Γ be the same sets as defined in Section 4 and used from then on. We assume that both Ω and Γ are K-vector spaces (K = arbitrary field) and recall the definition of the shift operators σ_Ω and σ_Γ on Ω and Γ (see (3.10)). We denote both shift operators by z but ignore, until later, the $K[z]$-module structure on Ω and Γ.

By a constant (not necessarily linear) input/output map $f: \Omega \to \Gamma$ we shall mean any map f which commutes with the shift operators, that is,

$$f(z \cdot \omega) = z \cdot (f(\omega)).$$

Let us now formulate the general problem of this section:

(10.1) PROBLEM OF OBSERVABILITY. *Given an input/output map f, its canonical realization Σ, and an input sequence $\nu \in \Omega$ applied after $t = 0$. Determine the state x of Σ at $t = 0$ from the knowledge of the output sequence of Σ after $t = 0$.*

This problem cannot be solved in general! To see this, recall that the state set X_f of f may be viewed as a set of functions

$$\{f(\omega \circ \cdot)(1): \Omega \to K^p: \nu \mapsto f(\omega \circ \nu)(1)\}$$

since ω' is Nerode-equivalent to ω iff

$$f(\omega' \circ \cdot)(1) = f(\omega \circ \cdot)(1)$$

Giving $\nu \in \Omega$ and the corresponding output sequence amounts to giving various values of $f(\omega \circ \cdot)(1)$ (namely those corresponding to the sequences $\emptyset$, ν_r, $z\nu_r + \nu_{r-1}$, ..., ν, $z\nu$, $z^2\nu$, ...), and it may happen that these substitutions do not yield enough values of the function $f(\omega \circ \cdot)(1)$ to determine the function itself. This situation has been recognized for a long time in automata theory, where, in an almost self-explanatory terminology, one says that "Σ is initial-state determinable by an infinite multiple experiment (possibly infinitely many different ν's) but not necessarily by a single experiment (single ν chosen at will)." See MOORE [1956]. The problem is further complicated by the fact that it may make a difference whether or not we have a free choice of ν. KALMAN, FALB, and ARBIB [1969, Section 6.3)] give some related comments.

A further difficulty inherent in the preceding discussion is that the problem is posed on a purely set-theoretic level and does not lend itself to the introduction of more refined structural assumptions. We shall therefore reformulate the problem in such a way as to focus attention on determining those properties of the initial state which can be computed from the combined knowledge of the input and output sequence occurring after $t = 0$.

For simplicity, we shall fix the value of ν at 0 (no loss of generality, since f is not linear). Then the output sequence resulting from x after $t = 0$ is given simply as $f(\omega)$, where $x = [\omega]_f$.

We shall use the circumflex to denote certain classes of functions from a set into the field K. For the moment, this class will be the class of all functions. Thus

$$\hat{\Gamma} = \{\text{all functions } \Gamma \to K\}.$$

An element $\hat{\gamma}$ of $\hat{\Gamma}$ is simply a "rule" (in practice, a computing algorithm) which assigns to each possible output sequence γ in Γ

a number in the field K. If γ resulted from the state $x = [\omega]_f$, then

$$\hat{\gamma}(\gamma) = \hat{\gamma}(f(\omega)) = (\hat{\gamma}\circ f)(\omega)$$

gives the value of a certain function in $\hat{\Omega}$ and, by definition of the state, also the value of a certain function in $\hat{X}$. This suggests the

(10.2) DEFINITION. An element $\hat{x} \in \hat{X}$ is an observable costate iff there is a $\hat{\gamma}_{\hat{x}} \in \hat{\Gamma}$ such that we have identically for all $\omega \in \Omega$

$$\hat{x}([\omega]_f) = \hat{\gamma}_{\hat{x}}(f(\omega)).$$

In other words, no matter what the initial state $x = [\omega]_f$ is, the value of $\hat{x}$ at x can always be determined by applying the rule $\hat{\gamma}_{\hat{x}}$ to the output sequence $f(\omega)$ resulting from x. Note, carefully, that this definition subsumes (i) a fixed choice of the class of functions denoted by the circumflex, and (ii) a fixed input sequence after $t = 0$ (here $\nu = 0$). For certain purposes, it may be necessary to generalize the definition in various ways [KALMAN 1970 a], but here we wish to avoid all unessential complications.

According to Definition (10.2), we shall see that a system is completely observable iff every costate is observable. This agrees with the point of view adopted earlier (see Section 4) in an ad-hoc fashion. Also, the vague requirement to "determine x" used in

(10.1) is now replaced by a precise notion which can be manipulated (via the actual definition of the circumflex) to express limitations on the algorithms that we may apply to the output sequence of the system.

The requirement "every costate is observable" can be often replaced by a much simpler one. For instance, if X is a vector space, it is enough to know that "every linear costate is observable" or even just that "every element of some dual basis is an observable costate"; if X is an algebraic variety, it is natural to interpret "complete observability" as "every element of the coordinate ring of X is an observable costate" [KALMAN 1970a].

We can now carry out a straightforward "dualization" of the setup involved in the definition of the input/output map $f\colon\ \Omega \to \Gamma$. First, we adopt (again with respect to a fixed interpretation of the circumflex):

(10.3) DEFINITION. The dual of an input/output map $f\colon\ \Omega \to \Gamma$ is the map

$$\hat{f}\colon\ \hat{\Gamma} \to \hat{\Omega}\colon\ \hat{\gamma} \mapsto \hat{\gamma} \circ f$$

Note that $\hat{f}$ is well-defined, since the circumflex means the class of all functions.

As to the next step, we wish to prove that constancy is inherited under dualization. To do this, wo have to induce a definition of the shift operator on $\hat{\Gamma}$ and $\hat{\Omega}$. The only possible definitions are the obvious ones:

$$\sigma_{\hat{\Gamma}}\colon\ \hat{\Gamma} \to \hat{\Gamma}\colon\ \hat{\gamma} \mapsto [\sigma_{\hat{\Gamma}}\hat{\gamma}\colon\ \gamma \mapsto \hat{\gamma}(\sigma_{\Gamma}\gamma)];$$

$$\sigma_{\hat{\Omega}}\colon\ \hat{\Omega} \to \hat{\Omega}\colon\ \hat{\omega} \mapsto [\sigma_{\hat{\Omega}}\hat{\omega}\colon\ \omega \mapsto \hat{\omega}(\sigma_{\Omega}\omega)].$$

Both of these new shift operators will be denoted by z^{-1}. The reason for this notation will become clear later.

Now it is easy to verify:

(10.4) PROPOSITION. If f is constant, so is $\hat{f}$.

PROOF. We apply the definitions in suitable sequence:

$$\begin{aligned}
\hat{f}(z^{-1}\cdot\hat{\gamma})(\omega) &= (z^{-1}\cdot\hat{\gamma})(f(\omega)) && (\text{def. of } \hat{f}),\\
&= \hat{\gamma}(z\cdot f(\omega)) && (\text{def. of } \sigma_{\hat{\Gamma}}),\\
&= \hat{\gamma}(f(z\cdot\omega)) && (f \text{ is constant}),\\
&= \hat{f}(\hat{\gamma})(z\cdot\omega) && (\text{def. of } \hat{f}),\\
&= (z^{-1}\cdot\hat{f}(\hat{\gamma}))(\omega) && (\text{def. of } \sigma_{\hat{\Omega}}),
\end{aligned}$$

and so we see that $\hat{f}$ commutes with z whenever f does. □

At this stage, we cannot as yet view $\hat{f}$ as the input/output map of a dynamical system because concatenation is not yet defined on $\hat{\Gamma}$, and therefore $\hat{\Gamma}$ is not yet a properly defined "input set". In other words, it is necessary to check that the notion of time is also inherited under dualization. In general, this does not appear to be possible without some strong limitation on the class $\hat{\Gamma}$. Here we shall look only at the simplest

(10.5) HYPOTHESIS. *Every function* $\hat{\gamma}$ *in* $\hat{\Gamma}$ *satisfies the finiteness condition: There is an integer* $|\hat{\gamma}|$ (*dependent on* $\hat{\gamma}$) *such that for all* $\gamma, \delta \in \Gamma$ *the condition*

$$\gamma_k = \delta_k, \; k = 1, \ldots, |\hat{\gamma}|$$

implies

$$\hat{\gamma}(\gamma) = \hat{\gamma}(\delta).$$

In other words, we assume that the value of each $\hat{\gamma}$ at γ is uniquely determined by some *finite portion* of the output sequence γ.

Assuming (10.5), it is immediate that $\hat{\Gamma}$ admits a concatenation multiplication which corresponds (at least intuitively) to the usual one defined on Ω:

(10.6) $$\hat{\gamma} \circ \hat{\delta} = z^{-|\hat{\delta}|} \cdot \hat{\gamma} + \hat{\delta}.$$

We can now prove the expected theorem, which may be regarded as the precise form of the "duality" principle:

(10.7) THEOREM. *Let* f *be an arbitrary constant input/output map and* $\hat{f}$ *its dual. Suppose further that* (10.5) *holds. Then each observable costate of* f (*relative to* $\hat{\Gamma}$ *satisfying* (10.5)) *may be viewed as a reachable state of* $\hat{f}$, *and conversely.*

PROOF. First we determine the Nerode equivalence classes on $\hat{\Gamma}$ induced by $\hat{f}$. By definition

$$\hat{\delta} \in (\hat{\gamma})_{\hat{f}} \text{ iff } \hat{f}(\hat{\delta} \circ \hat{\epsilon}) = \hat{f}(\hat{\gamma} \circ \hat{\epsilon})$$

for all $\hat{\epsilon} \in \hat{\Gamma}$. Now $\hat{f}$ is linear (!); in fact, direct use of the definition of $\hat{f}$ and (10.6) gives

$$\hat{\delta} \in (\hat{\gamma})_{\hat{f}} \quad \text{iff} \quad (\hat{\gamma} \circ f)(\omega) = (\hat{\delta} \circ f)(\omega), \; \omega \in \Omega.$$

So $\hat{\gamma} \circ f$ and $\hat{\delta} \circ f$ are equal as elements of X: they define the same observable costate. In fancier language, the assignment

(10.8) $\quad d\colon X_{\hat{f}} \to \hat{X}_f\colon \; (\hat{\gamma})_{\hat{f}} \mapsto \hat{\gamma} \circ f$

is well defined and constitutes a bijection between the reachable states of $\hat{f}$ and those costates of f which are observable relative to the function class $\hat{\ }$. □

Thus (10.5) is a sufficient condition for the duality principle to hold. However, the fact that the canonical realization of $\hat{f}$ is completely reachable is not quite the same as saying that the canonical realization of f is completely observable because the latter <u>depends</u> on the choice of $\hat{\Gamma}$ and therefore is not an intrinsic property of f. Moreover, Theorem (10.7) does not give any indication how "big" $X_{\hat{f}}$ is and it may certainly happen that the observability problem for f is much more difficult than the reachability problem. These matters will be illustrated later by some examples.

Now we deduce the original form of the duality principle from Theorem (10.7). The essential point is that (10.5) holds automatically as a result of linearity.

New definition of the function class: let the circumflex denote the class of all <u>K-linear</u> functions. (All the underlying sets with the K-vector spaces, so the definition makes sense.)

The following facts are well known:

(10.9) PROPOSITION. Let $*$ denote duality in the sense of K-vector spaces. Then:

$$\hat{\Gamma} \triangleq (K^p[[z^{-1}]])^* = K^p[z^{-1}],$$
$$\hat{\Omega} \triangleq (K^m[z])^* = K^m[[z]].$$

Now we can state the

(10.10) MAIN THEOREM. Suppose f is K-linear, constant, finite-dimensional. Suppose further that $\hat{}$ means K-linear duality. Then:

(i) $\hat{f}$ is K-linear and constant, that is, a $K[z^{-1}]$-homomorphism (and therefore written as f^*) and finite-dimensional.

(ii) The reachable states of f^* are isomorphic with the K-linear dual of X_f; hence every costate of X_f is observable.

PROOF. The fact that Γ is K-linear implies, by (10.3), that $\hat{f}$ is K-linear; the constancy of f always implies that of $\hat{f}$, by Proposition (10.4). (Caution: $\hat{f}$ is not the K[z]-linear dual of the K[z]-homomorphism f, and the construction given here cannot be simplified. See Remark (4.26A).)

To prove the second part, we note that by Proposition (10.9) Hypothesis (10.5) holds and thus $\hat{f} = f^*$ is a well-defined input/output map of a dynamical system. We must prove that the reachable states of f^* are isomorphic with X_f^*, the K-linear dual of X_f. This amounts to proving that the K-vector space of functions

$$x \mapsto \hat{\gamma}(h_f(x),\ h_f(z\cdot x),\ \dots\ )$$

is isomorphic with the K-vector space X_f^*. It suffices to prove that the K-vector space generated by the K-linear functions

$$(10.11)\qquad \{\lambda:\ x \mapsto [h_f(z^i \cdot x)]_j,\quad i = 0, 1, \ldots \text{ and } j = 1, \ldots, m\}$$

is isomorphic with X_f^*. Suppose that, for fixed x, every $\lambda(x) = 0$. Then $x = 0$, by definition of the Nerode equivalence relation induced by f (recall here the discussion from Section 3). Since X_f is finite-dimensional by hypothesis, it follows from this property of the functions $\{\lambda\}$ that they generate X_f^*. Obviously, $\dim X_f^* = \dim X_f$, so that everything is proved. □

In other terms, the fact that f = K[z]-homomorphism <u>together with the appropriate definition of</u> ^ implies that

$$\hat{f}:\ K^p[z^{-1}] \to K^m[[z]]$$

is a $K[z^{-1}]$-homomorphism. Since (10.5) holds, we can interpret $\hat{f}$ in a system-theoretic way, as follows: the output of the dual system at $t = -k$ due to input $\hat{\gamma}$ is given by the assignment

$$\hat{\gamma} \mapsto \hat{f}(\hat{\gamma})(-k),$$

which is a linear function defined on the k-th term of the input sequence. In fact, we have

$$\begin{aligned}\hat{\gamma}(\gamma) &= \hat{f}(\hat{\gamma})(\omega),\\ &= (\hat{\gamma}\circ f)(\omega),\\ &= \sum_k(\hat{f}(\hat{\gamma})(-k))(\omega_k).\end{aligned}$$

(10.12) REMARK. It is essentially a consequence of Proposition (10.9) that $\hat{f}$ turns out to be the same kind of algebraic object as f. Note, however, that

> <u>under duality the input and output terminals are interchanged and</u> t <u>is replaced by</u> $-t$ (<u>hence</u> z <u>by</u> z^{-1}).

In terms of the pictorial definition of a system, this statement simply amounts to "reversing the directions of the arrows", which is the "right" way to define duality in the most general mathematical context, namely in category theory. We would expect that the duality principles of system theory will eventually become a part of this very general duality theory. This has not happened yet because the correct categories to be considered in the study of dynamical systems have not yet been determined. It is likely that eventually many different categories will have to be looked at in studying dynamical problems.

We shall now present an example which should help to interpret the previous results. We emphasize, however, that the theory sketched here is still in a very rudimentary form.

(10.13) EXAMPLE. Consider the system Σ defined by

$$x(t+1) = 2x(t) + u(t), \quad y(t) = x(t), \quad t \in \underline{\underline{Z}};$$

$$y(t) = \begin{cases} 0 & \text{if } 0 \leqq x(t) < 1/2, \\ 1 & \text{if } 1/2 \leqq x(t) < 1, \end{cases}$$

with $X = U = Y = \underline{\underline{R}}$ mod 1, i.e., the interval $[0, 1)$. (1 is to be thought of as identified with 0.) We let $u(t) = 0$. We view x through its binary representation

$$x = \sum_{k=0}^{\infty} \xi_k(x) 2^{-k}, \qquad \xi_k(x) = 0 \text{ or } 1.$$

It is clear from the definition of the system that the output sequence due to any x is precisely

$$\gamma_x = (\xi_1(x), \xi_2(x), \dots).$$

If x is irrational, infinitely many terms are needed to identify it. Consequently, the x's are isomorphic with the Nerode equivalence classes induced by f_Σ. So Σ cannot be reduced.

Relative to "^ = functions", every costate of f_Σ is observable, provided that Hypothesis (10.5) is <u>not</u> satisfied. If it is, then only those costates defined on fixed-length rationals are observable (more precisely, these are functions which depend only on a fixed finite subset of the $\xi_k(x)$'s). Thus: <u>either $\hat{f}$ does not define a dynamical system or not all costates are observable</u>.

Now let us replace the set $[0, 1)$ by its intersection with the rationals. It is clear that there is now a <u>finite</u> algorithm for determining x: we simply apply the results of partial realization theory of the previous section. (We take $K = \underline{\underline{Z}}_2$ and the problem is to express x from $(\xi_1(x), \dots, \xi_2(x)0$ as a ratio of polynomials in $\underline{\underline{Z}}_2[2]$--which is always possible since each x is rational.) However, x is not "effectively computable" in the strict sense since there is no way of knowing when the algorithm has stopped. In other words, given an arbitrary costate $\hat{x}$ there exists no <u>fixed</u> rule $\hat{\gamma}_{\hat{x}}$ such that the application of $\hat{\gamma}_{\hat{x}}$ to γ_x gives $\hat{x}(x)$ for all x. On the other hand, substituting into $\hat{x}$ the results of the partial-realization algorithm will give an approximation to the value of $\hat{x}(x)$ which always converges in a finite (but a priori unknown) number of steps as more values of the output sequence are observed. In short, the costate-determination algorithm has certain pseudo-random elements in it and therefore cannot be described through the machinery of deterministic dynamical systems. (Is there some relation here to the conceptual difficulties of Quantum Mechanics?)

11. HISTORICAL COMMENTS

It is not an exaggeration to say that the entire theory of linear, constant (and here, discrete-time) dynamical systems can be viewed as a systematic development of the equivalent algebraic conditions (2.8) and (2.15).

Of course, the use of modules (over $K[z]$) to study a constant square matrix (see (4.13)) has been "standard" since the 1920's under the influence of E. NOETHER and especially after the publication of the Modern Algebra of VAN DER WAERDEN. Condition (2.15), by itself, must be also quite old. For instance, GANTMAKHER [1959, Vol. 1, p. 203] attributes to KRYLOV [1931] the idea of computing the characteristic polynomial of a square matrix A by choosing a random vector b and computing successively $b, Ab, A^2b, \ldots$ until linear dependence is obtained, which yields the coefficients of $\det(zI - A)$. (The method will succeed iff X_A is cyclic with generator g.) However, the merger of (4.13) with (2.15), which is the essential idea in the algebraic theory of linear systems, was done explicitly first in KALMAN [1965b].

We shall direct our remarks here mainly to the history of conditions (2.8) and (2.15) as related to controllability. See also earlier comments in KALMAN [1960c, pp. 481, 483, 484] and in KALMAN, HO, and NARENDRA [1963, pp. 210-212]. We will have to bear in mind that the development of modern control theory cannot be separated from the development of the concept of controllability; moreover, the technological problems of the 1950's and even earlier had a major influence on the genesis of mathematical ideas (just as the latter have led to many new technological applications of control in the 1960's).

R. E. Kalman

The writer developed the mathematical definition of controllability with applications to control theory, during the first part of 1959. (Unpublished course notes at Johns Hopkins University, 1958/59.) These first definitions were in the form of (2.17) and (2.3). Formal presentations of the results were made in Mexico City (September, 1959, see KALMAN [1960b]), University of California at Berkeley (April, 1969, see KALMAN [1960d]), and Moskva (June, 1960, see KALMAN [1960c]), and in scientific lectures on many other concurrent occasions in the U.S. As far as the writer is aware, a conscious and explicit definition of controllability which combines a control-theoretic wording with a precise mathematical criterion was first given in the above references. There are of course many instances of similar ideas arising in related contexts. Perhaps the comments below can be used as the starting point of a more detailed examination of the situation in a seminar in the history of ideas.

The following is the chain of the writer's own ideas culminating in the publications mentioned above:

(1) In KALMAN [1954] it is pointed out (using transform methods) that continuous-time linear systems can be controlled by a linear discrete-time (sampled-data) controller in finite time.*

----

*It is sometimes claimed in the mathematical literature of optimal control theory that this cannot be done with a linear system. This is false; the correct statement is "cannot be done with a linear controller producing control functions which are continuous (and not merely piecewise continuous!) in time." Such a restriction is completely irrelevant from the technological point of view. As a matter of fact, computer-controlled systems have been proposed and built for many years on the basis of linear, time-optimal control.

(2) Transposing the result of KALMAN [1954] from transfer functions to state variables, an algorithm was sketched for the solution of the discrete-time time-optimal control of systems with bounded control and linear continuous-time dynamics. [KALMAN, 1957]

(3) As a popularization of the results of the preceding work, the same technique was applied to give a general method for the design of linear sampled-data systems by KALMAN and BERTRAM [1958].

Some background comments concerning these papers are appropriate:

(1) The ideas and method presented in KALMAN [1954] descend directly from earlier (and very well known) engineering research on time-optimal control. (The main references in KALMAN [1954] are: McDONALD [1950], HOPKIN [1951], BOGNER and KAZDA [1954], as well as a research report included in KALMAN [1955].) Although the results of KALMAN [1954] on linear time-optimal control were considered to be new when published, it became clear later that similar ideas were at least implicit in OLDENBOURG and SARTORIUS [1951, §90, p. 219] and in TSYPKIN's work in the early 1950's. The engineering idea of nonlinear time-optimal control goes back, at least, to DOLL [1943] and to OLDENBURGER in 1944, although the latter's work was unfortunately not widely known before 1957. During the same time, there was much interest in the same problems in other countries; see, for instance, FELDBAUM [1953] and UTTLEY and HAMMOND [1953]. Mathematical work in these problems probably began with BUSHAW's dissertation [1952] in which, to quote from KALMAN [1955, before equation (40)], " ... [it was] rigorously proved that the intuition which led to the formulation of the [engineering] theory [quoted above] was indeed correct." TSIEN's survey [1954] contains a lengthy account of this state

of affairs and was ready by many. We emphasize: none of this extensive literature contains even a hint of the algebraic considerations related to controllability.

(2-3) The critical insight gained and recorded in KALMAN [1957] is the following: the solution of the discrete-time time-optimal control problem is equivalent to expressing the state as a linear combination of a certain vector sequence (related to control and dynamics) with coefficients bounded by 1 in absolute value, the coefficients being the values of the optimal control sequence. The linear independence of the first n vectors of the sequence guarantees that every point in a neighborhood of zero can be moved to the origin in at most n steps (hence the terminology of "complete controllability"); and the condition for this is identical with (2.17) (stated in KALMAN [1957] and KALMAN and BERTRAM [1958] only for the case $\det F \neq 0$ and $m = 1$). A thorough discussion of these matters is found in KALMAN [1960c; see especially Theorem I, p. 485]. A serious conceptual error in KALMAN [1957] occurred, however, in that complete controllability was not <u>assumed</u>, as a hypothesis for the existence of time-optimal control law, but an attempt was made to show that the controllability is almost always complete [Lemma 1]. In fact, this lemma <u>is</u> true, with a small technical modification in the condition. Only much later did it become clear (see the discussion of Theorem D in the Introduction), however, that a dynamical system is <u>always</u> completely controllable (in the nonconstant case, completely reachable) if it is derived from an external description. It was this difficulty, very mysterious in 1957, which led to the development of a formal machinery for the definition of controllability during the next two years. The changing point of view is already apparent in KALMAN and BERTRAM [1958]; the unpublished paper promised there was delayed precisely because the algebraic machinery to prove Theorem D was out of reach in 1957-8. Consult also the findings of the bibliographer RUDOLF [1969].

IN SUMMARY: under the stimulation of the engineering problems of minimal-time optimal control, the researches begun by KALMAN [1954, 1957] and KALMAN and BERTRAM [1958] eventually evolved into what has come to be called the mathematical theory of controllability (of linear systems).

Beginning about 1955, and stimulated by the same engineering problems, PONTRYAGIN and his school in the USSR developed their mathematical theory of optimal control around the celebrated "Maximum Principle". (They were well aware of the survey of TSIEN [1954] mentioned above, and referenced it both in English and in the Russian translation of 1956.) We now know that any theory of control, regardless of its particular mathematical style, must contain ingredients related to controllability. So it is interesting to examine how explicitly the controllability condition appears in the work of PONTRYAGIN and related research.

GAMKRELIDZE [1957, §2; 1958 §1, §2] calls the time optimal control problem associated with the system

(11.1) $dx/dt = Ax + bu(t)$

"nondegenerate" iff b is not contained in a proper A-invariant subspace of R^n. He notes immediately that this is equivalent to

(11.2) $\det (b, Ab, \ldots, A^{n-1}b) \neq 0$

(i.e., the special case of (2.8) for $m = 1$). He then proves: in the "degenerate" case the problem either reduces to a simpler one or the motion cannot be influenced by the control function $u(\cdot)$. All this is very close to an explicit definition of controllability. However, in discussing the general case $m > 1$, GAMKRELIDZE [1958, §3, Section 1] defines "nondegeneracy" of the system

(11.3) $dx/dt = Ax + Bu(t)$

as the condition

(11.4) $\det (b_i, Ab_i, \ldots, A^{n-1}b_i) \neq 0$ for every column $b_i \in B$,

but he does not show that this generalized condition of "nondegeneracy" for (11.3) inherits the interesting characterization proved for "nondegeneracy" in the case of (11.1). In fact, condition (11.4) is much too strong to prove this; the correct condition is (2.8), that is, complete controllability. In other words, in GAMKRELIDZE's work (11.4) plays the role of a technical condition for eliminating "degeneracy" (actually, lack of uniqueness) from a particular optimal control problem and is not explicitly related to the more basic notion of complete controllability. Neither GAMKRELIDZE nor PONTRYAGIN [1958] give an interpretation of (11.4) as a property of the dynamical system (11.3), but employ (11.4) only in relation to the particular problem of time-optimal control. See also KALMAN [1960c, p. 484]. A similar point of view is taken by LASALLE [1960]; he calls a dynamical system (11.3) satisfying (2.8) "proper" but then goes on to require (11.4) (to assure the uniqueness of the time-optimal controls) and calls such systems "normal".

The assumption of some kind of "nondegeneracy" condition was apparently unavoidable in the early phases of research on the time-optimal control problem. For example, ROSE [1953, pp. 39-58] examines this problem for (11.1); by defining "nondegeneracy" [p. 41] by a condition equivalent ot (11.2), he obtains most of GAMKRELIDZE's results in the special case when A has real eigenvalues [Theorem 12]. ROSE uses determinants closely related to the now familiar lemmas in controllability theory but he, too, fails to formulate controllability as a concept independent of the time-optimal control problem.

A similar situation exists in the calculus of variations. The so-called Caratheodory classes (after CARATHEODORY [1933]) correspond to a kind of classification of controllability properties of nonconstant systems. In fact, the standard notion of a normal family of extremals of the calculus of variations is closely related to condition (11.4), suitably generalized via (2.5) to nonconstant systems.* Normality is used in the calculus of variations mainly as a "nondegeneracy" condition.

It is important to note that the "nondegeneracy" conditions employed in optimal control and the calculus of variations play mainly the role of eliminating annoying technicalities and simplifying proofs.

----

*The use of the word "normal" by LaSALLE [1960] for (11.4) is only accidentally coincident with the earlier use of the "normal" in the calculus of variations.

With suitable formulation, however, the basic results of time-optimal control theory continue to hold <u>without</u> the assumption of complete controllability. The same is not true, however, of the four kinds of theorems mentioned in the Intorduction, and therefore these results are more relevant to the story of controllability than the time-optimal control discussed above.

There is a considerable body of literature relevant to controllability theory which is quite independent of control theory. For instance, the treatment of a reachability condition in partial differential equations goes back at least to CHOW [1940] but perhaps it is fairer to attribute it to Caratheodory's well-known approach to entropy via the nonintegrability condition. The current status of these ideas as related to controllability is reviewed by WEISS [1969, Section 9]. An independent and very explicit study of reachability is due to ROXIN [1960]; unfortunately, his examples were purely geometric and therefore the paper did not help in clarifying the celebrated condition (2.8). The Wronskian determinant of the classical theory of ordinary differential equations with variable coefficients also has intersections with controllability theory, as pointed out recently with considerable success by SILVERMAN [1966]. Many problems in control theory were misunderstood or even incorrectly solved before the advent of controllability theory. Some of these are mentioned in KALMAN [1963b, Section 9]. For relations with automata theory, see ARBIB [1965].

Let us conclude by stating the writer's own current position as to the significance of controllability as a subject in mathematics:

(1) Controllability is basically an algebraic concept. (This claim applies of course also to the nonlinear controllability results obtained via the Pfaffian method.)

(2) The historical development of controllability was heavily influenced by the interest prevailing in the 1950's in optimal control theory. Ultimately, however, controllability is seen as a relatively minor component of that theory.

(3) Controllability as a conceptual tool is indispensable in the discussion of the relationship between transfer functions and differential equations and in questions relating to the four theorems of the Introduction.

(4) The chief current problem in controllability theory is the extension to more elaborate algebraic structures.

For a survey of the historical background of observability, which would take us too far afield here, the reader should consult KALMAN [1969b].

12. REFERENCES

Section A: General References

M. A. ARBIB

[1965] A common framework for automata theory and control theory, SIAM J. Contr., 3:206-222.

C. W. CURTIS and I. REINER

[1962] Representation Theory of Finite Groups and Associative Algebras, Interscience-Wiley.

E. M. DAY and A. D. WALLACE

[1967] Multiplication induced in the state space of an act, Math. System Theory, 1:305-314.

C. A. DESOER and P. VARAIYA

[1967] The minimal realization of a nonanticipative impulse response matrix, SIAM J. Appl. Math., 15:754-764.

E. G. GILBERT

[1963] Controllability and observability in multivariable control systems, SIAM J. Control, 1:128-151.

B. L. HO and R. E. KALMAN

[1966] Effective construction of linear state-variable models from input/output functions, Regelungstechnik, 14:545-548.

[1969] The realization of linear, constant input/output maps, I. Complete realizations, SIAM J. Contr., to appear.

S. T. HU

[1965] Elements of Modern Algebra, Holden-Day.

R. E. KALMAN

[1960a] A new approach to linear filtering and prediction problems, J. Basic Engr. (Trans. ASME), 82D:35-45.

[1960b] Contributions to the theory of optimal control, Bol. Soc. Mat. Mexicana, 5:102-119.

[1960c] On the general theory of control systems, Proc. 1st IFAC Congress, Moscow; Butterworths, London.

[1962] Canonical structure of linear dynamical systems, Proc. Nat. Acad. of Sci. (USA), 48:596-600.

[1963a] New methods in Wiener filtering theory, Proc. 1st Symp. on Engineering Applications of Random Function Theory and Probability, Purdue University, November 1960, pp 270-388, Wiley. (Abridged from RIAS Technical Report 61-1.)

[1963b] Mathematical description of linear dynamical systems, SIAM J. Contr., 1:152-192.

[1965a] Irreducible realizations and the degree of a rational matrix, SIAM J. Contr., 13:520-544.

[1965b] Algebraic structure of linear dynamical systems. I. The Module of Σ, Proc. Nat. Acad. Sci. (USA), 54:1503-1508.

[1967] Algebraic aspects of the theory of dynamical systems, in Differential Equations and Dynamical Systems, J. K. Hale and J. P. LaSalle (eds.), pp. 133-146, Academic Press.

[1969a] On multilinear machines, J. Comp. and System Sci., to appear.

[1969b] Dynamic Prediction and Filtering Theory, Springer, to appear.

[1969c] On partial realizations of a linear input/output map, Guillemin Anniversary Volume, Holt, Winston and Rinehart.

[1970a] Observability in multilinear systems, to appear.

[1970b] The realization of linear, constant, input/output maps. II. Partial realizations, SIAM J. Control, to appear.

R. E. KALMAN and R. S. BUCY

[1961] New results in linear prediction and filtering theory, J. Basic Engr. (Trans. ASME, Ser. D), 83D:95-100.

R. E. KALMAN, P. L. FALB and M. A. ARBIB

[1969] Topics in Mathematical System Theory, McGraw-Hill.

R. E. KALMAN, Y. C. HO and K. NARENDRA

[1963] Controllability of linear dynamical systems, Contr. to Diff. Equations, 1:189-213.

C. E. LANGENHOP

[1964] On the stabilization of linear systems, Proc. Am. Math. Soc., 15:735-742.

[1965] Algebra, Addison-Wesley.

S. MAC LANE

[1963] Homology, Springer.

L. A. MARKUS

[1965] Controllability of nonlinear processes, SIAM J. Control, 3:78-90.

E. F. MOORE

[1956] Gedanken-experiments on sequential machines, in Automata Studies, C. E. Shannon and J. McCarthy (eds.), pp. 129-153, Princeton University Press.

P. MUTH

[1899] Theorie und Anwendung der Elementartheiler, Teubner, Leipzig.

A. NERODE

[1958] Linear automaton transformations, Proc. Amer. Math. Soc., 9:541-544.

L. SILVERMAN

[1966] Representation and realization of time-variable linear systems, Doctoral dissertation, Columbia University.

L. M. SILVERMAN and H. E. MEADOWS

[1969] Equivalent realizations of linear systems, SIAM J. Control, to appear.

H. WEBER

[1898] Lehrbuch der Algebra, Vol. 1, 2nd Edition, reprinted by Chelsea, New York.

L. WEISS

[1969] Lectures on Controllability and Observability, C.I.M.E. Seminar.

L. WEISS and R. E. KALMAN

[1965] Contributions to linear system theory, Intern. J. Engr. Sci., 3:141-171.

W. M. WONHAM

[1967] On pole assignment in multi-input controllable linear systems, IEEE Trans. Auto. Contr., AC-12:600-665.

A. M. YAGLOM

[1962] An Introduction to the Theory of Stationary Random Functions, Prentice-Hall.

D. C. YOULA

[1966] The synthesis of linear dynamical systems from prescribed weighting patterns, SIAM J. Appl. Math., 14:527-549.

D. C. YOULA and P. TISSI

[1966] n-port synthesis via reactance extraction, Part I, IEEE Intern. Convention Record.

O. ZARISKI and P. SAMUEL

[1958] Commutative Algebra, Vol. 1, Van Nostrand.

Section B: References for Section 11

M. A. ARBIB

[1965] A common framework for automata theory and control theory, SIAM. J. Contr., 3:206-222.

I. BOGNER and L. F. KAZDA

[1954] An investigation of the switching criteria for higher order contactor servomechanisms, Trans. AIEE, 73 II:118-127.

D. W. BUSHAW

[1952] Differential equations with a discontinuous forcing term, doctoral dissertation, Princeton University.

C. CARATHEODORY

[1933] Über die Einteilung der Variationsprobleme von Lagrange nach Klassen, Comm. Mat. Helv., 5:1-19.

W. L. CHOW

[1940] Über Systeme von linearen partiellen Differentialgleichungen erster Ordnung, Math. Annalen, :98-105.

H. G. DOLL

[1943] Automatic control system for vehicles, US Patent 2,463,362.

A. A. FELDBAUM

[1953] Avtomatika i Telemekhanika, 14:712-728.

R. V. GAMKRELIDZE

[1957] On the theory of optimal processes in linear systems (in Russian), Dokl. Akad. Nauk SSSR, 116:9-11.

[1958] The theory of optimal processes in linear systems (in Russian), Izvestia Akad. Nauk SSSR, 2:449-474.

F. R. GANTMAKHER

[1959] The Theory of Matrices, 2 vols., Chelsea.

A. M. HOPKIN

[1951] A phase-plane approach to the compensation of saturating servomechanisms, Trans. AIEE, 70:631-639.

R. E. KALMAN

[1954] Discussion of a paper by Bergen and Ragazzini, Trans. AIEE, 73 II: 245-246.

[1955] Analysis and design principles of second and higher-order saturating servomechanisms, Trans. AIEE, 74 II:294-310.

[1957] Optimal nonlinear control of saturating systems by intermittent control, IRE WESCON Convention Record, 1 IV:130-135.

[1960b] Contributions to the theory of optimal control, Bol. Soc. Mat. Mexicana, 5:102-119.

[1960c] On the general theory of control systems, Proc. 1st IFAC Congress, Moscow; Butterworths, London.

[1960d] Lecture notes on control system theory (by M. Athans and G. Lendaris), Univ. of Calif. at Berkeley.

[1963b] Mathematical description of linear dynamical systems, SIAM J. Contr., 1:152-192.

[1965b] Algebraic structure of linear dynamical systems. I. The Module of Σ, Proc. Nat. Acad. Sci. (USA), 54:1503-1508.

[1969b] Dynamic Prediction and Filtering Theory, Springer, to appear.

R. E. KALMAN and J. E BERTRAM

[1958] General synthesis procedure for computer control of single and multi-loop linear systems, Trans, AIEE, 77 III:602-609.

R. E. KALMAN, Y. C. HO and K. NARENDRA

[1963] Controllability of linear dynamical systems, Contr. to Diff. Equations, 1:189-213.

A. N. KRYLOV

[1931] On the numerical solution of the equation by which the frequency of small oscillations is determined in technical problems (in Russian), Izv. Akad. Nauk SSSR Ser. Fix.-Mat., 4:491-539.

J. P. LaSALLE

[1960] The time-optimal control problem, Contr. Nonlinear Oscillations, Vol. 5, Princeton Univ. Press.

D. C. McDONALD

[1950] Nonlinear techniques for improving servo performance, Proc. Nat. Electronics Conf. (USA), 6:400-421.

R. C. OLDENBOURG and H. SARTORIUS

[1951] Dynamik selbstättiger Regelungen, 2nd edition, Oldenbourg, Munchen.

R. OLDENBURGER

[1957] Optimum nonlinear control, Trans. ASME, 79:527-546.

[1966] Optimal and Self-Optimizing Control, MIT Press.

L. S. PONTRYAGIN

[1958] Optimal control processes (in Russian), Uspekhi Mat. Nauk, 14:3-20.

N. J. ROSE

[1953] Theoretical aspects of limit control, Report 459, Stevens Institute of Tech., Hoboken, N.J.

E. ROXIN

[1960] Reachable zones in autonomous differential systems, Bol. Soc. Mat. Mexicana, 5:125-135.

K. E. RUDOLF

[1969] On some unpublished works of R. E. Kalman, not to be unpublished.

H. S. TSIEN

[1954] Engineering Cybernetics, McGraw-Hill.

A. M. UTTLEY and P. H. HAMMOND

[1953] The stabilization of on-off controlled servomechanisms, in Automatic and Manual Control, Academic Press.

L. WEISS

[1969] Lectures on Controllability and Observability, C.I.M.E. Seminar

CENTRO INTERNAZIONALE MATEMATICO ESTIVO

(C. I. M. E.)

R. KULIKOWSKI

CONTROLLABILITY AND OPTIMUM CONTROL

Corso tenuto a Sasso Marconi dal 1 al 9 luglio 1968

CONTROLLABILITY AND OPTIMUM CONTROL

by
R. Kulikowski
(Polish Academy of Sciences, Warszawa)

1. Introduction and Statement of the Problem

The intuitive notion of controllability of dynamic systems had been used for many years by control-engineers. In the case of linear stable systems, described by eqs.

(1) $$\dot{x} = Ax + Bu \quad ,$$

(2) $$y = Cx \quad ,$$

where

x = n-dimensional state-vector,

u = r-dimensional control-vector,

y = p-dimensional output-vector,

A, B, C - matrices of the dimensions $n \times n$, $n \times r$, $p \times n$, respectively,

that notion has been formulated strictly by Kalman (see Ref. [2, 3]) .

According to Kalman the system S, described by (1), (2) is controllable if and only if: given that the system is in state x_o at time $t=0$, then for some finite time $T > 0$ there is a control $u(t)$, $t \in [0, T]$ such that $x(T) = 0$.

He has proved also that S is controllable if and only if the column vectors of the matrix

(3) $$Q \triangleq [B, AB, \ldots, A^{n-1} B]' ,$$

where $B, \ldots, A^{n-1} B$ denote columns of Q, span the state space R^n of S. That means that the rank of Q is n or that among nr columns of Q there is a set of n linearly independent vectors which constitute a basis for R^n.

In control theory the notion of optimal (time) control is also being used. The system S is optimally (time) controlled if there exists the control function $u(t) \in U$, where U is an admissible control-set (usually a convex r-dimensional polyhedron) such that the time T required for the transition of state variables from the given initial state x_o to the given final state $x(T)$ (e.g. $x(T) = 0$) is minimum.

Using the well known "maximum principle" of Pontryagin (see Ref. [8]), or an equivalent optimization technique, it is possible to formulate the necessary condition of optimality for system S. That condition can be written in the form of linear differential equations and it can be used for derivation of optimum control function $u(t)$. As shown by Pontryagin and others (Ref. [8]) the necessary condition of optimality becomes also sufficient when the system is controllable. Besides, the optimum control exists and it is unique.

Then the notion of controllability has an important theoretical and practical value. It determines conditions under which the optimum control of a linear system is generally possible.

Roughly speaking a controllable system is the system in which the state variables can be driven to any position with a finite value of performance measure and an optimally controllable system is the system in which the system in which the state variables can be driven to the given position with a minimum value of performance measure. In the case of the linear systems and optimization time as the performance measure the controllable system can be optimally controlled. Is that also true if we take as performance measure the other possible performance criteria? A simple example will show that the answer to that question is negative.

Consider the first-order system described by the eq.

$$dx/dt = u(t), \quad x(0) = 0$$

R. Kulikowski

Assume that the control force u(t) is generated by the controller having finite energy, what can be written as $u \in L^2[0, T]$. As the performance measure we assume the so called "square error":

$$E(u) = \| y - x \|^2 = \int_0^T \left[y(t) - \int_0^T u(\tau)\, d\tau \right]^2 dt,$$

where y(t) is a given square-integrable function.

The necessary condition of optimality requires that

$$y(t) = \int_0^t u(\tau)\, d\tau \quad , \quad t \in [0, T] .$$

When that condition holds the functional E(u) attains its lower bound: $E(\bar{u}) = 0$.

Assuming that $y(t) = 1(t)$ (a unit step) we find easily that the optimum value of u(t) is Dirac's $\delta(t)$ function. However, that function is not square-integrable and the optimum solution of our optimization problem does not exist despite the fact that the finite evergy solutions "approximating" $\delta(t)$ exist.

Obviously the difficulties of the present type arise in the case when some sufficient conditions of optimality (such as the generalized Weierstrass theorem) for nonlinear functionals do not hold .

In the present paper we shall deal with a "well defined" class of optimalization problems for which the optimum control exists and an optimally controlled system is also a controllable system (at least in certain of state space).

It will be usegull to introduce the following definition.

We shall say that a system is optimally controllable if the system state-variables can be derive to any (given) position with minimum value of the performance measure and the system constraints are not violated.

We shall strive to find conditions under which the optimum solutions of concrete optimization problems exist and to determine the regions of controllability in the space of system and constraint parameters.

We shall be interested mainly in complex optimalization problems including energy and amplitude inequality constraints, constraints of phase coordinates, nonlinear systems etc.

Dealing with that class of problems it is convenient to employ a few simple notions of functional analysis. Then the main optimization problem which is investigated in the present paper can be formulated as follows. Let a nonlinear functional $F(x)$, $x \in X$ (where X is a Banach space) be given. Find the point $x_o \in X$ such that

$$F(x_o) = \max_{x \in \Omega} F(x) \tag{4}$$

where $\Omega \subset X$ is the set of feasible solutions, specified by the condition

$$G(x) \geq 0 \tag{5}$$

where G is a nonlinear operator ($G : X \rightarrow Z$ and Z is another Banach space). It is also assumed that F and G are concave and strongly differentiable.

In the most practical applications the assumption that X and Z are Banach spaces (i.e. the linear, normed and complete spaces). holds. In particular we shall deal with the space of all continuous functions $C[0, T]$, and with the space of all p-power integrable functions $L^p[0, T]$, $p \geq 1$. In these cases the element x is a function of time $(x(t))$ and since G is an operator the element $z = G(x)$, $z \in Z$ is another function of time $(z(t))$. The inequality (5) means that the values $z(t)$ are nonnegative for all $t \in [0, T]$ - in the case $Z \equiv C$, and for almost all $t \in [0, T]$ - in the case $Z \equiv L^p$.

We shall write also $z_1 \geqslant z_2$ if $z_1 - z_2 \geqslant 0$. The inequalities defined in such a way have the known properties of inequalities for numbers (i.e. one can multiply them by non negative numbers, sum them up and pass to the limits on both sides). Generally, one can say that Z includes a convex, closed cone (i.e. the set closed with respect to the multiplication by real nonnegative numbers). The closed cone induces in Z a partial--order relation denoted by the sign $\geq$.

One should observe that the constraint of the form

(6) $$a(t) \leq x(t) \leq b(t) ,$$

where $a(t)$, $b(t)$ = given functions, can be represented in the general form (5) if we introduce the operators:

$$G_1(x) = x - a ,$$

$$G_2(x) = b - x ,$$

and write

$$G(x) = < G_1(x), G_2(x) > .$$

Using that notation we assume that for an ordered pair $z = < z_1, z_2 >$ the inequality $z \geq 0$ is equivalent to the pair of inequalities $z_1 \geq 0$, $z_2 \geq 0$. In the similar way for the n ordered set

$$z = < z_1, z_2, \ldots, z_n >$$

the inequality $z \geq 0$ is equivalent to $z_i \geq 0$, $i = 1, \ldots, n$.

The space of all ordered pairs $< z_1, z_2 >$, where $z_1 \in Z_1$ and $z_2 \in Z_2$ is called the Cartesian product of Z_1 and Z_2 . It is denoted by $Z_1 \times Z_2$. In the above mentioned example the domain and range of operators G_1, G_2 is X (i.e. $G_1 : X \to X$, $G_2 : X \to X$) and $G(x) \in Z = X \times X$.

It should be noted that if for a function x two inequality constraints hold

$$G_o(x) \geq 0 \quad \text{and} \quad G_o(x) \leq 0 ,$$

then the equality constraint $G_o(x) = 0$ is also valid. Then the equality constraint is a special case of (5), where $G(x) = < G_o(x), - G_o(x) >$.

As can be seen the constraint (5) may also express the equality or inequality constraints imposed on the state and output coordinates of a dynamic system (1), (2).

In order to obtain an explicite relation of that type one may integrate eq. (1) and express the state and output coordinates as a Volterra operator of control vector.

The functional F may describe the control cost. If, for example, x denotes a vector-function, with components : $u_1(t), u_2(t), \ldots, u_r(t)$, the energy cost of control forces is proportional to

$$\int_0^T \sum_{i=1}^r [u_i(t)]^2 \, dt ,$$

and

$$F(x) = - \int_0^T \sum_{i=1}^r [u_i(t)]^2 \, dt . \tag{7}$$

In the present case the optimization time T is fixed. As can be seen later it will be also possible to solve problems in which T is minimized.

Before we formulate the sufficient and necessary conditions of optimality it is necessary to introduce the notion of weak and strong differentials.

Then a method based on the notion of Lagrange-functional will be presented. General idea of that method is based on the paper of L. Hurwicz [1] dealing with nonlinear programming in topological spaces. As shown in Ref. [4, 5, 7] the method of Lagrange-functionals has been found to be very useful in the solution of optimization problems with operator inequality constraints.

2. Differentiation of nonlinear operators and functionals

Let $G(x)$ be a nonlinear operator $G: X \to Y$, where X, Y are Banach spaces, and x, h be elements of X. The operator G is called weakly differentiable if the limit

(1) $$\lim_{\gamma \to 0} \frac{1}{\gamma}\{G(x + \gamma h) - G(x)\} = dG(x, h) ,$$

where γ is a number, exists.

The differential $dG(x, h)$ is a homogeneous operator with respect to h, i.e. $dG(x, \gamma h) = \gamma . dG(x, h)$. When $dG(x, h)$ exists and is continuous at x it is also an additive operator (see Ref. [9]).

As an example find the weak differential of the operator

(2) $$G(x) = \int_0^T K[t, \tau , x(\tau)] d\tau, \qquad x \in C[0, T] ,$$

where the functions $K[t, \tau, x]$, $K'_x[t, \tau, x] = \frac{\partial K}{\partial x}[t, \tau, x]$ are continuous with respect to $t, \tau \in [0, T]$ and $x \in X$.

We get formally

(3) $$d G(x, h) = \frac{dG(x + \gamma h)}{d\gamma}\bigg|_{\gamma = 0} = \int_0^T K'_x[t, \tau , x(\tau)] \, h(\tau) \, d\tau .$$

Since the function K is uniformly continuous the differentiation under the integral sign is admissible.

Besides the weak differential the strong (or Frechet) differential is also being used. If at the point $x \in X$ we have

(4) $$G(x + h) - G(x) = dG(x, h) + \Omega(x, h) ,$$

where $dG(x, h)$ is a linear operator with respect to $h \in X$ and

$$\lim_{\|h\| \to 0} \frac{\|\Omega(x, h)\|}{\|h\|} = 0 ;$$

we call $dG(x, h)$ the strong differential of $G(x)$.

When $dG(x,h)$ is a linear operator acting from the space X into Y, it is also an element of the Banach space of all linear operators acting from X into Y and it is called the derivative of $G(x)$ at the point x. Denoting that operator by $G'(x)$ we can write

(5) $$d(G(x, h) = G'(x)(h)$$

In that formula $G'(x)$ is an operator acting on the element $h \in X$.

As follows from the definition of strong differential if it exists it is equal to the weak differential . It is possible to show (see Ref. [9]) that the continuity and existence of weak derivative $G'(x)$ at the vicinity of x is sufficient for the existence of strong derivative and that these two derivatives are equal.

Now we shall consider relations between the differentials and derivatives of differentiable functionals. Denote the weak linear differential of the functional $F(x)$, $x \in \Omega \subset X$ by $dF(x, h)$. Since $dF(x, h)$ is a linear functional with respect to h one can write $dF(x,h) = (y,h)$, where y is an element of the space X^* adjoint to X. That element can be treated also as a result of an operation i.e. $y = f(x)$, $f: X \to X^*$.

Consider, for example, the functional

$$F(x) = \int_0^T K[x(\tau), \tau] \, d\tau , \quad x \in L^p [0, T] .$$

where

$$K[x,\tau] , \quad \frac{d}{dx} K[x,\tau] = K'_x [x, \tau] , \quad \tau \in [0, T] ,$$

are continuous functions.

We get by (3)

$$dF(x, h) = \int_0^T K'_x [x,\tau] h(\tau) d\tau = \int_0^T y(\tau) \, h(\tau) \, d\tau .$$

Since $x, h \in L^p[0, T]$ then $y(\tau) \in L^q[0, T]$, where $p^{-1} + q^{-1} = 1$, and the operator

(6) $$y = f(x) = K'_x[x, \tau]$$

is acting from $L^p[0, T]$ into $L^q[0, T]$

The operator f is being called the gradient of the functional F:

$$f(x) = \text{grad } F(x)$$

In the case of the finite dimensional space R^n the gradient is usually denoted by $\nabla F(x)$.

3. Lagrange functionals

By Lagrange functional we call the expression

(1) $$\Phi(x, \lambda) = F(x) + \lambda[G(x)] ,$$

where λ is a linear functional defined over the space Z (of the range of operator $G(x)$).

The form of the functional λ depends on the space Z. In practical applications Z is usually the space $L^p[0, T]$ or $C[0, T]$. For these spaces there one knows the general forms of linear functionals ([6]). However we shall be mainly interested in the non-negative functionals. We call a functional $\lambda(z)$ non-negative if for all the non-negative $z \geq 0$ it is non-negative, i.e. $\lambda(z) \geq 0$.

The general form of a linear non-negative functional in $L^p[0, T]$ space is

(2) $$\lambda(z) = \int_0^T z(t)\, \lambda(t)\, dt ,$$

where $\lambda(t)$ is almost every-where non-negative function ($\lambda(t) \geq 0$ for almost every $t \in [0, T]$) and $\lambda(t) \in L^q$, $q = p/(p-1)$.

The non-negative functionals over the space $C[0, T]$ assume the following form

$$\lambda(z) = \int_0^T z(t) \, d\lambda(t) , \tag{3}$$

where $\lambda(t)$ is a non-decreasing, continuous from the right side, function having bounded variation.

In the general case the constraint $G(x) \geq 0$ consists of several components, i.e.

$$G_i(x) \geq 0, \quad i = 1, \ldots, n ,$$

where $G_i : X \to Z_i$, $i = 1, \ldots, n$, and the functional λ is defined over the Carthesian product

$$Z_1 \times Z_2 \times \ldots \times Z_n .$$

It consists of n components i.e.

$$\lambda = \langle \lambda_1, \lambda_2, \ldots, \lambda_n \rangle$$

where λ_i denotes the linear functional over Z_i.

When we write $\lambda \geq 0$ it means that all λ_i are non-negative.

The Lagrange functional in the present case can be written

$$\Phi(x, \lambda) = F(x) + \sum_{i=1}^{n} \lambda_i [G_i(x)] , \qquad \text{where} \quad \lambda_i \in Z_i^* .$$

Since $F(x)$, $\lambda[G(x)]$ are concave functionals of x, $\Phi(x, \lambda)$ is a concave functional of x and a linear functional of λ.

For concave differentiable functions $F(x)$ the following inequality holds

$$F(x_2) \leq F(x_1) + dF(x_1, x_2 - x_1) . \tag{4}$$

It follows from the definition of the differential and the definition of a concave function:

(5) $$F[\alpha x_1 + (1-\alpha)x_2] \geq \alpha F(x_1) + (1-\alpha)F(x_2),$$

$0 < \alpha < 1$, x_1, x_2 - arbitrary points of the domain of $F(x)$

It can be easily proved that the inequality (4) holds also for concave and differentiable functionals.

4. Sufficient condition of optimality

Theorem 1. Let the functional F and the operator G be concave and strongly differentiable over X. If there exists such a $\bar{x} \in X$ and a linear non-negative functional $\bar{\lambda}$ that for every $x \in X$

(1) $$d_x \Phi(\bar{x}, \bar{\lambda}, x) = 0 ,$$

(2) $$G(\bar{x}) > 0 ,$$

(3) $$\bar{\lambda}[G(\bar{x})] = 0,$$

(4) $$\bar{\lambda} \geq 0$$

then the functional $F(x)$ attains at $\bar{x}$ the maximum value subject to the constraint $G(\bar{x}) \geq 0$.

Proof [1]:

Since F, λG are concave functionals we have

$$F(x_o) \leq F(\bar{x}) + dF(\bar{x}, x_o - \bar{x}) ,$$

$$\bar{\lambda}[G(x_o)] \leq \bar{\lambda}[G(\bar{x})] + \bar{\lambda} dG(\bar{x}, x_o - \bar{x}) ,$$

1) The proof of a similar theorem is given in Ref. [1] .

where x_o = arbitrary element of X. Summing up these inequalities one gets

$$F(x_o) + \bar{\lambda}[G(x_o)] \leq F(\bar{x}) + \bar{\lambda}[G(\bar{x})] + d_x \Phi[(\bar{x}, \bar{\lambda}), x_o - x] .$$

Taking into account (1) and (3) one obtains

$$F(\bar{x}) \geq F(x_o) + \bar{\lambda}[G(x_o)] .$$

Since $\bar{\lambda} \geq 0$ and $G(x_o) \geq 0$ we get $F(\bar{x}) \geq F(x_o)$.
Since x_o is arbitrary the theorem has been proved.

Remark 1 .

When the constraint $x \geq 0$ is distinghished explicitly i.e. when

$$G(x) = \langle G_1(x), x \rangle ,$$

where $G: X \to Z$, $Z = Z_1 \times X$,
one can write

$$\Phi(x, \lambda) = \Phi_1(x, \lambda_1) + \lambda_2(x) ,$$

where $\lambda = \langle \lambda_1, \lambda_2 \rangle$.

Then the conditions (1) - (4) can be written in the following equivalent form :

(5) $$d_x \Phi_1 [(\bar{x}, \bar{\lambda}_1), \bar{x}] = 0$$

(6) $$d_x \Phi_1 [(\bar{x}, \bar{\lambda}_1), x] \leq 0, \quad \text{for every} \quad x \geq 0 ,$$

(7) $$d_\lambda \Phi_1 [(\bar{x}, \bar{\lambda}_1), \bar{\lambda}_1] = 0 ,$$

(8) $$d_\lambda \Phi_1 [(\bar{x}, \bar{\lambda}_1), \lambda_1] \geq 0 \quad \text{for every} \quad \lambda_1 \geq 0$$

Indeed, the condition (1) can be written

(9) $$d_x \Phi [(\bar{x}, \bar{\lambda}), x] = d_x \Phi_1 [(\bar{x}, \bar{\lambda}_1), x] + \bar{\lambda}_2(x) = 0$$

Since $\bar{\lambda}_2(x) = 0$, for $x = \bar{x}$ one gets (5). According to (4) for every $x \geq 0$ the inequality $\bar{\lambda}_2(x) \geq 0$ holds. Then by (9) one gets (6).

Since $\bar{\lambda}_1[G_1(\bar{x})] = d_{\lambda_1}\Phi[(\bar{x}, \bar{\lambda}_1), \lambda_1]$ then (3) can be written in the form of (7). Taking into account that $G_1(\bar{x}) \geq 0$ if and only if $\lambda_1[G_1(\bar{x})] \geq 0$ for every $\lambda_1 \geq 0$, the relation (2) can be written in the form of (8).

Remark 2:

It should be noted that when F and G are convex and there exists such a pair $(\bar{x}, \bar{\lambda})$, that for every $x \in X$ the conditions

(10) $$d_x\Phi(\bar{x}, \bar{\lambda}; x) = 0,$$

(11) $$G(\bar{x}) \leq 0,$$

(12) $$\bar{\lambda}[G(\bar{x})] = 0,$$

(13) $$\bar{\lambda} \geq 0,$$

hold then $F(x)$ attains at $\bar{x}$ its minimum value; subject to $G(x) \leq 0$. The formulae (5) - (8) preserve their form, except (6) and (8), which change the in equality signs, i.e.:

(14) $$d_x \Phi_1[(\bar{x}, \bar{\lambda}), x] \geq 0, \quad \text{for every} \quad x \geq 0,$$

(15) $$d_\lambda \Phi_1[(\bar{x}, \bar{\lambda}_1), \lambda_1] \leq 0, \quad \text{for every} \quad \lambda \geq 0,$$

Remark 3.

In the case when the explicit form of the functional $\Phi(x, \lambda)$ is given the optimality conditions including differentials can be replaced by conditions in the gradient form.

Let for example, $F(x)$ be the integral operator

$$F(x) = \int_0^T K[x(\tau), \tau] \, d\tau$$

having the gradient

$$f[x, \tau] = K'_x[x(\tau), \tau],$$

(see (6) of sec. 2). Assume that $G(x)$ be a differentiable operator and $G: X \to L^p[0, T]$.

Taking into account the general form of linear functionals in $L^p[0, T]$ (see (2) of sec. 3) one can write

$$G(x) = \int_0^T \lambda(\tau)\, G[x]\, d\tau, \quad \lambda(\tau) \in L^q[0, T] .$$

Denote the gradient of that functional by $\lambda(\tau)\, g[x,\tau]$, i.e.

$$\text{grad}\, \lambda\, [G(x)] = \lambda(\tau)\, g[x, \tau],$$

and the gradient of $\Phi(x, \lambda)$ by $\varphi[x, \lambda, \tau]$.

For a fixed $x = \bar{x}$

$$\varphi[\bar{x}, \lambda, \tau] = f[\bar{x}, \tau] + \lambda(\tau)\, g[\bar{x}, \tau]$$

is a function of $\tau \in [0, T]$ and one can write

$$d_x \Phi[\bar{x}, \bar{\lambda}\,;\, x] = (\text{grad}_x\, \Phi(\bar{x}, \bar{\lambda}), x) = \int_0^T \varphi[\bar{x}, \bar{\lambda}, \tau]\, x(\tau)\, d\tau .$$

The conditions (1), (10) reduce to the requirement that almost everywhere

$$\text{grad}_x\, \Phi(\bar{x}, \bar{\lambda}) = \varphi[\bar{x}, \bar{\lambda}, \tau] = 0 . \tag{16}$$

In a similar way the conditions (5), (6) are equivalent to the following :

I) if $\text{grad}_x\, \Phi_1(\bar{x}, \bar{\lambda}) < 0$ for almost all $\tau \in P \subset [0, T]$, then $\bar{x}(\tau) = 0$ almost every-where in P ,

II) if $\bar{x}(\tau) > 0$ for almost all $\tau \in R \subset [0, T]$, then $\text{grad}_x\, \Phi_1(\bar{x}, \bar{\lambda}) = 0$ almost every-where in R ,

where P, R = arbitrary sets of positive measure of the interval $[0, T]$.

The conditions (3), (4) (and (7), (8)) can be also written in the gradient form :

III) if $\text{grad}_{\lambda} \Phi(\bar{x}, \bar{\lambda}) = G(\bar{x}) > 0$ for almost all $\tau \in P \subset [0, T]$ then
$\bar{\lambda}(\tau) = 0$ almost every-where in P,

IV) if $\bar{\lambda}(\tau) > 0$ for almost all $\tau \in R \subset [0, T]$, then
$\text{grad}_{\lambda} \Phi(\bar{x}, \bar{\lambda}) = 0$ almost every-where in R.

Conditions III-IV have a simple physical interpretation: at the points where the constraints are not active the Lagrange-function vanishes and when the Lagrange - function is positive the constraints must be active.

It is also possible to write down the gradient form of conditions of optimality in the case of minimalization of $F(x)$.

When the operator $G(x)$ is acting into the space of continuous functions, i.e. $G : X \to C[0, T]$, one can write

$$\lambda [G(x)] = \int_0^T G[x] \, d\lambda(\tau) ,$$

where $\lambda(\tau)$ is a non-increasing function, what can be also written as $d\lambda(\tau) \geq 0$ (that notation means that the increase of $\lambda(\tau)$ in the arbitrary small vicinity of τ is non-negative). When $\lambda(\tau)$ is differentiable $d\lambda(\tau) = \lambda'(\tau)\, d\tau$, and the above condition can be written $\lambda'(\tau) \geq 0$. The last inequality is equivalent to $d\lambda(\tau) \geq 0$ if the differentiation is defined in the generalized sense (e.g. as the Schwartz's distribution). In that case $\lambda'(\tau)$ can also include Dirac's $\delta(t)$ - functions at the discontinuity-points of $\lambda(\tau)$.

The condition (3) in the present case can be written as

$$G[x]\, d\bar{\lambda}(\tau) = 0, \quad \tau \in [0, T]$$

or

I) if $\text{grad}_{\lambda} \Phi(\bar{x}, \bar{\lambda}) = G[\bar{x}] > 0$ for $\tau \in P \subset [0, T]$, then $d\lambda(\tau) = 0$ ($\lambda(\tau) = \text{const}$) at the point τ or vicinity of τ,

II) if $d\lambda(\tau) > 0$, $\tau \in P$ then $G[\bar{x}] = 0$, $\tau \in P$.

It is also possible to write the gradient form of optimality conditions for the remaining formulae (1) - (4), (5) - (8), (10) - (13). We shall leave that as an exercise for the readers.

The conditions (5) - (8) can be called the quasi-saddle-point conditions and they can be treated as a generalization of well known Kuhn-Tucker differential conditions of optimality, which were formulated originally for nonlinear functions in finite-dimensional spaces R^n .

5. Necessary conditions of optimality.

When an optimization problem is being solved it is also important to know that the functions x(t) which do not satisfy the conditions (1)-(4) or (5) - (8) can not be optimal. That problem requires to prove that the conditions (1)-(4) are also necessary for optimum. In order to prove the necessary conditions we shall impose certain regularity conditions on F and G but shall not assume any more that F and G are concave.

We shall call $x_o \in X$ a regular point of G if for every admissible variation x at the point x_o , which is defined by the condition

$$(1) \qquad G[x_o] + dG(x_o, x) \geq 0 ,$$

there exists e curve emanating from x_o, tangent to x_o and lying in the set of admissible solutions Ω . By a curve in the Banach space we understand, generally speaking, a function ψ of real variable s with the range in X , i.e. $\psi : R \to X$. According to definition that function should satisfy the following conditions:

$$(2) \qquad \psi(s) \in \Omega , \quad \psi(0) = x_o, \qquad d\psi(0, 1) = x .$$

We shall show now, that if $\bar{x}$ is a maximalizing point for F(x) and a regular point of G then for each admissible variation x, i.e.

$$(3) \qquad G(x) + dG(\bar{x}, x) \geq 0 ,$$

we get a non-positive increase of F, i.e.

$$(4) \qquad -dF(\bar{x}, x) \geq 0.$$

Indeed, the real function

$$(5) \qquad f(s) = F[\psi(s)]$$

attains the maximum value for $\psi(0) = \bar{x}$, i.e. for $s = 0$. Then $df(0, 1) \leq 0$. On the other hand, according to the differentiation rule of compound. function (5) and formulae (2) we get

$$df(0, 1) = dF[\psi(0), d\psi(0, 1)] = dF(\bar{x}, x).$$

Then $dF(\bar{x}, x) \leq 0$ and (4) has been proved.

Introducing the notation

$$l_1(x) = -dF(\bar{x}, x), \quad g(x) = G(\bar{x}) + dG(\bar{x}, x)$$

the obtained result can be written as:

$$(6) \qquad \text{if } g(x) \geq 0 \quad \text{then } l_1(x) \geq 0.$$

The next step in our reasoning consists in showing that there exists such a functional $\bar{\lambda} \geq 0$, which will ensure the relation

$$l_1(x) = \bar{\lambda}[L_1(x)],$$

where

$$(7) \qquad L_1(x) = dG(\bar{x}, x).$$

The main obstacle in showing that is the nonlinearity of $g(x)$, which consists of linear term $L_1(x)$ and the additive term $G(\bar{x})$.

In order to overcome that obstacle an auxiliary operator

$$(8) \qquad L < s, x > = < s, sG(\bar{x}) + dG(\bar{x}, x) >,$$

where

$$L : R \times X \rightarrow R \times Z, \quad s \in R,$$

can be introduced. it can be proved that $L < s, x >$ is a linear operator, i.e. for real numbers α_1, α_2 and $s_1, s_2 \in R$, $x_1, x_2 \in X$ one gets:

$$L < \alpha_1 s_1 + \alpha_2 s_2, \alpha_1 x_1 + \alpha_2 x_2 > =$$

$$= \alpha_1 \; L < s_1, x_1 > + \alpha_2 L < s_2, x_2 > .$$

Since the functional $l_1(x)$ can be also treated as defined over $R \times X$ then the following notation

$$l < s, x > = -dF(\bar{x}, x)$$

can be introduced.

Now we should check whether the condition

$$L <s, x> \geq 0 \text{ implies } l <s, x> \geq 0.$$

Observe that this condition can be written as

(9) $$s \geq 0 \quad \text{and} \quad sG(\bar{x}) + dG(\bar{x}, x) \geq 0$$

Assuming first of all that $s > 0$ and divining (9) by s we get

$$G(\bar{x}) + dG(\bar{x}, x/s) \geq 0 .$$

Taking into account (4) we obtain $-dF(\bar{x}, x/s) \geq 0$ and $-dF(\bar{x}, x) \geq 0$ or

(10) $$l < s, x> \geq 0 .$$

Then the functional l is nonnegative on the set P of pairs $< s, x >$ which satisfly conditions $s > 0$, $sG(\bar{x}) + dG(\bar{x}, x) \geq 0$.

Each point $<0, x>$ for which the relation

(11) $$dG(\bar{x}, x) \geq 0$$

holds can be treated as a limit of the sequence of points from P. That sequence can be constructed as follows.

Take on element x_o with the property that

$$\frac{1}{n} G(\bar{x}) + dG(\bar{x}, x_o/n) \geq 0 \tag{12}$$

Summing up (11) and (12) one gets

$$\frac{1}{n} G(\bar{x}) + dG(\bar{x}, x_o/n + x) \geq 0$$

Then

$$< \frac{1}{n}, \frac{1}{n} x_o + x > \in P \quad . \text{ Since}$$

$$\lim_{n \to \infty} < \frac{1}{n}, \frac{1}{n} x_o + x > = < 0, x >$$

then (9) implies (10) for all $s \geq 0$.

Introducing the notation $<s, x> = w$, $R \times X = W$ the obtained relation can be written as

$$\text{if} \quad L(w) \geq 0 \quad \text{then} \quad l(w) \geq 0 , \tag{13}$$

where

L = linear operator, $L : W \to V = R \times Z$,

l = linear functional over W (an element of the space W^*, adjoint to W).

We can now return back to the main problem which is the existence of nonnegative functional $\bar{v}^*$ satisfying the relation

$$l(w) = \bar{v}^*[L(w)] . \tag{14}$$

To solve that problem we shall need a generalization for Banach spaces of the well known Farkas lemma.

First of all it is necessary to introduce a few additional notions.

We shall denote by Q a set of all functionals of w which can be represented in the form

$$w^*(w) = v^*[L(w)] , \quad v^* \geq 0 , \tag{15}$$

where

$$v^* \in V^*, \quad V = R \times Z .$$

The sequence of functionals $w_k^* \in W^*$ is called weakly convergent to $w^* \in W^*$, if the sequence $w_k^*(w)$, $k = 1, 2, \ldots$ converges to $w^*(w)$ for each $w \in W$.

The set of functionals $Q \subset W^*$ is called weakly closed if together with each weakly convergent sequence $w_k^* \to w$ it contains also their limits, i.e. $w_k^*, w \in Q$.

The generalized Farkas lemma (called the lemma of Minkowski and Farkas) states that if Q is a weakly closed set then it contains each functional w^*. In other words there exists such $\bar{v}^* \geq 0$ that (14) holds (see Ref. [1]).

Now we are able to formulate and prove the following necessary conditions of optimality.

<u>Theorem 2.</u>

Let

1.) at the point $\bar{x}$ the functional $F(x)$ attains the conditional maximum, subject to the constraint $G(x) \geq 0$,

2) the point $\bar{x}$ is a regular point $G(x)$,

3) the set Q is weakly closed.

Then there exists such a functional $\bar{\lambda} \geq 0$ that the quasi-saddle-point conditions (1) - (4) hold.

<u>Proof</u> [1].

It was shown already that there exists such a functional $\bar{v}^* \geq 0$, that

$$1 < s, x > = \bar{v}^*(L < s, x >) . \tag{16}$$

Taking into account (8) one can write

1) The proof of a similar theorem is given in Ref. [1].

$$\bar{v}^*\langle s, x\rangle = \mu s + \bar{\lambda}[sG(\bar{x}) + dG(\bar{x}, x)] ,$$

where u is a real number and $\bar{\lambda} \in Z^*$.

The condition $\bar{v}^* \geq 0$ can be written in the form

(17) $$\mu \geq 0, \bar{\lambda} \geq 0$$

Hence (4) follows .

The relation (16) can be rewritten as

(18) $$-dF(\bar{x}, x) = \mu s + \bar{\lambda}[sG(\bar{x}) + dG(\bar{x}, x)] .$$

Setting $s = 0$ one gets from (18)

$$dF(\bar{x}, x) + \bar{\lambda}[dG(\bar{x}, x)] = d_x \Phi[(\bar{x}, \bar{\lambda}), x] = 0$$

and

$$\text{grad}_x \Phi[\bar{x}, \bar{\lambda}] = 0 .$$

Then the relation (1) follows.

Setting $x = 0$, $s = 1$ one gets $\mu + \bar{\lambda}[G(\bar{x})] = 0$, what by (17) yields $\bar{\lambda}[G(\bar{x})] = 0$ and (3). The condition (2) must hold by assumption, Q.E.D.

The theorem 2 is valid under two regularity conditions. In the examples which are considered below the operator G(x) assumes the form $G(x) = A(x) + a$, where A(x) is a linear operator over X , and a - given element of X. One can easily check that G(x) is a regular operator. For each admissible variation x, defined by

$$G(\bar{x}) + dG(\bar{x}, x) = A(\bar{x}) + a + A(x) \geq 0 ,$$

and the function $\psi(s) = \bar{x} + sx$, one obtains

$$\psi(s) = \bar{x} + sx \in \Omega , \quad \psi(0) = \bar{x}, \quad d\psi \quad (0, 1) = x$$

and the curve ψ is lying within the set of admissible solutions.

It is possible to check that the same property have the operators G(x.), which consist of several components (constraints) $A_i(x) + a_i$, A_i-

linear, $a_i \in X$, $i = 1, \ldots, n$.

In the case of nonlinear operators $G(x)$, which are acting into the space of continuous functions $(Z = C[0, T])$ or into the space of measurable, bounded functions $(Z = L^{\infty}[0, T])$ the regularity condition boils down (see in that respect Ref. [7]) to the existence of admissible variation x, such that the function $z(t)$:

$$z(t) = G[\bar{x}] + dG(\bar{x}, x),$$

is positive. In the case of the space of continuous functions it means that

$$\min_{0 \leq t \leq T} z(t) > 0,$$

and in the case of the space $L^{\infty}[0, T]$ it means that the minimum of $z(t)$ is positive almost every-where.

As an example consider the inequality

(20) $$G(x) = a(t) - \int_0^t \varphi[x(\tau)]\, d\tau \geq 0, \quad 0 \leq t \leq T,$$

where

$\varphi(x)$ is increasing and having continuous derivative $\varphi'(x)$; $a(t)$ = = nonnegative, continuous function. That operator is regular . Indeed, since $G(x) \geq 0$ and

$$dG(\bar{x}, x) = - \int_0^t \varphi'[x(\tau)]\, x(\tau)\, d\tau, \quad \varphi'(x) > 0,$$

one can set $x(\tau) = -1$ and obtain $z(t) > 0$.

As shown in Ref. [7] in order to check whether the regularity condition 3. of theorem 2 holds it is sufficient to show that there exists such an element $w^* \in W$, that

(21) $$L(w^*) > 0 .$$

For example, in the case of the operator (20) one obtains

$$L(w) = < s, \int_0^t \varphi'\left[\bar{x}(\tau)\right] x(\tau) d\tau + s\left\{a(t) - \int_0^t \varphi\left[\bar{x}(\tau)\right] d\tau\right\} >, \quad 0 \leq t \leq T.$$

Since $a(t) - \int_0^t \varphi[\bar{x}(\tau)]\, d\tau \geq 0$, then it can be easily proved that there exists such a pair $< s^*, x^* >$, $s^* > 0$ and $x^*(t) > 0$, $t \in [0, T]$, that $L(w^*) > 0$.

It should be noted that the theorem 2 generalizes certain theorem of variational calculus in Banach spaces and, in particular, the following theorem of Lusternik.

Theorem 3.

Let the functionals F, H be strongly differentiable at the point $\bar{x} \in X$ and $\|\text{grad } H(\bar{x})\| > 0$. If $\bar{x}$ is a conditional extremum point of $F(x)$, subject to the constraint $H(x) = c$, where $c = H(\bar{x})$, then

(22) $$\text{grad } F(\bar{x}) = \mu \text{ grad } H(\bar{x}),$$

and μ is a number.

The proof of that theorem is given in Ref. [9].

6. Examples of optimum control problems.

Consider a linear dynamic system, shown in Fig. 1; having one controlled input $u(t)$ and $n+1$ outputs, which are described by Volterra operators:

(1) $$y_i(t) = \int_0^t k_i(t, \tau) u(\tau)\, d\tau,$$

where

$k_i(t,\tau)$ - linearly independent transient functions of the system,

$i = 0, 1, \ldots, n$.

A typical optimization problem can be formulated as follows: Find the function $u(t) \in L^p [0, T]$, which minimalizes

$$\|u\|_p = \left(\int_0^T |u(t)|^p \, dt \right)^{1/p}, \quad p \geq 1 \tag{2}$$

subject to the constraints

$$y_i(T) = \int_0^T k_i(T, \tau) \, u(\tau) \, d\tau = x_i, \quad i = 0, 1, \ldots, n \tag{3}$$

where T, x_i - given real numers, $k_i(T, \tau) \in L^q [0, T]$.

In other words, it is required to minimalize the control cost (2) for the given outputs x_i attained at the time $t = T$.

In certain cases additional conditions of the form

$$\underline{M} \leq u(t) \leq \overline{M}, \tag{4}$$

($\underline{M}$, $\overline{M}$ = given numbers) or

$$\mathcal{Y}_j(t) = \int_0^t k_j(t, \tau) \, u(\tau) \, d\tau \leq x_j(t), \quad j = 0, 1, \ldots, n \tag{5}$$

($x_j(t)$ = given time functions) are being imposed.

The constraint (5) is called "restriction of phase coordinates".

There exist, of course, many known optimization techniques, such as: variational calculus, maximum principle, dynamic programming etc. which can be applied for the solution of the optimization problems formulated above. In the present section we should like to demonstrate that the method based on theorems 1, 2 of sections 4, 5, is very convenient for the solution of problems including restriction of phase coordinates.

Instead of dealing with a general n-dimensional system we shall confine our analysis to a second order system, which is frequently encountered in the engineering practice (e.g. in servomechanisms etc.). The result of that analysis will be useful for the investigation of a class of controllability problems.

Example 1.

Consider a system described by the differential equation

(6) $$\frac{dy_1}{dt} = u(t), \quad \frac{dy_0}{dt} = y_1(t) ,$$

with zero initial conditions: $y_0(0) = y_1(0) = 0$, shown in Fig. 2.

It is required to find such a control function $u(t)$ which minimalizes the "energy cost":

(7) $$E(u) = \frac{1}{2} \int_0^T \left[u(\tau)\right]^2 d\tau,$$

subject to the constraints

(8) $$y_0(T) = \int_0^T (T - \tau) u(\tau) \, d\tau = x_0, \quad (x_0 > 0)$$

$$y_1(T) = \int_0^T u(\tau) \, d\tau = 0 .$$

The constraints (8), (9) mean that the deflection of the output coordinate of the system for $t=T$ is equal x_o and the corresponding velocity of that coordinate at $t = T$ is equal zero. The constraints of that kind are typical for operation of controlled motors and servomechanisms.

The Lagrangean of present problem is equal

(10) $$\Phi(u, \mu) = \frac{1}{2} \int_0^T \left[u(\tau)\right]^2 d\tau + \mu_1 \left\{ \int_0^T (T - \tau) u(\tau) \, d\tau - x_o \right\} + \mu_2 \int_0^T u(\tau) \, d\tau ,$$

where μ_1, μ_2 = Lagrange multipliers.

The necessary, and at the same time sufficient condition (due to the convexity of $E(u)$), of optimality according to theorem 3 becomes:

(11) $$\operatorname{grad}_u \Phi(u, \mu) = u(\tau) + \mu_1(T - \tau) + \mu_2 = 0 ,$$

where μ_1, μ_2 can be computed by (8), (9). We get then

$$\mu_1 = -\frac{12x_o}{T^3}, \quad \mu_2 = \frac{6x_o}{T^2},$$

and

(12) $$u(\tau) = \bar{u}(\tau) = \frac{6x_o}{T^2}\left(1 - \frac{2\tau}{T}\right), \quad 0 \leq \tau \leq T.$$

Now we can solve a more complicated problem when in addition to constraints (8), (9) the amplitude-constraints of control force:

(13) $$\begin{aligned} u(\tau) - M &\leq 0 \\ -M - u(\tau) &\leq 0 \end{aligned}$$

where M = given number, should be taken into account. These constraints are typical for the operation of electrical motors in servo-systems.

Since $u(\tau)$ is a square-integrable function the general form of the linear, non-negative functional $L(z)$ in the space $L^2[0, T]$ (according to (2) of section 3) becomes

$$L(z) = \int_0^T \lambda(\tau)\, z(\tau)\, d\tau ,$$

where the function $\lambda(\tau)$ is non-negative and square-integrable.

The Lagrangean of present problem becomes

$$\Phi(u, \mu, \lambda) = \frac{1}{2}\int_0^T [u(\tau)]^2 d\tau + \mu_1\left(\int_0^T (T - \tau)\, u(\tau)\, d\tau - x_o\right) +$$

$$+ \mu_2 \int_0^T u(\tau)\, d\tau + \int_0^T \lambda_1(\tau)\left[u(\tau) - M\right] d\tau + \int_0^T \lambda_2(\tau)\left[-M - u(\tau)\right] d\tau,$$

where $\lambda_1(\tau)$, $\lambda_2(\tau)$ = Lagrange functions.

The necessary and sufficient condition of optimality, according to formulae (10) - (13) of sec. 4, become

$$\operatorname{grad}_u \Phi(u, \mu, \lambda) = u(\tau) + \mu_1(T-\tau) + \mu_2 + \lambda_1(\tau) + \tag{15}$$
$$- \lambda_2(\tau) = 0,$$

$$u(\tau) - M \leq 0, \quad -M - u(\tau) \leq 0, \tag{16}$$

$$\int_0^T \lambda_1(\tau)\left[u(\tau) - M\right] d\tau = \int_0^T \lambda_2(\tau)\left[-M - u(\tau)\right] d\tau = 0 \tag{17}$$

$$\lambda_1(\tau) \geq 0, \quad \lambda_2(\tau) \geq 0, \tag{18}$$

where μ_1, μ_2 can be determined by (8), (9).

As can be seen these condition will hold for all τ, when $\lambda_1(\tau)$, $\lambda_2(\tau)$ become zero. That correspond to the nonactive constraints. When the constraints become active the functions $\lambda_1(\tau)$, $\lambda_2(\tau)$ should take such values which satisfy the condition (15). When the constraints are not active in the interval $[0, T]$ the optimum solution should be the same as (12).

Then the optimum solution becomes

$$u(t) = \begin{cases} M, & 0 \leq t \leq T_o \\ -\left[\mu_1(T-t) + \mu_2\right] = M\left[1 - \dfrac{2(t-T)}{T-2T_o}\right], & T_o \leq t \leq T-T_o \\ -M, & T - T_o \leq t \leq T \end{cases} \tag{19}$$

where T_o can be derived from the equation obtained by setting (19) into (8), which yields:

$$\int_0^T (T-t)\,\bar{u}(t)\,dt = M\left\{\int_0^T (T-t)dt + \int_{T_o}^{T-T_o} (T-t)\left[1 - \frac{2(t-T_o)}{T-2T_o}\right] dt\right.$$

$$- \int_{T_o - T}^{T} (T - t)\, dt \Big\} = \frac{M}{T - 2T_o} \left[\frac{T^3}{6} + \frac{2}{3} T^2 - T^2 T \right] = x_o .$$

Typical form of that solution has been shown in Fig. 3.

The correspondig optimum values of $\lambda_1(t)$, $\lambda_2(t)$ become:

$$\lambda_1(t) = \bar{\lambda}_1(t) = \begin{cases} -\mu_1(T - t) - \mu_2 - 2M = -4M \dfrac{(t - T_o)}{T - 2T_o}, & 0 \le t \le T_o \\ 0, \quad T_o \le t \le T \end{cases}$$

$$\lambda_2(t) = \bar{\lambda}_2(t) = \begin{cases} 0, \quad 0 \le t \le T - T_o \\ \mu_1(T - t) + \mu_2 - 2M = 4M \dfrac{t + T_o - T}{T - 2T_o}, & T - T_o \le t \le T \end{cases}$$

The plot of the relation $\frac{T_o}{T} = f\left[\frac{|x_o|}{MT^2}\right]$ has been shown in Fig. 4. It can be observed that the optimum solution, satisfying constraints (8), (9), (12) exists only if $|x_o| \le M(T/2)^2$. It means that when M is less that $4|x_o|/T^2$ it is impossible to attain the output x_o in time T using optimum (or non optimum) control. In other words, when $|x_o| < M(\frac{T}{2})^2$ the system is optimally controllable in the sense formulated in sec. 1.

Example 2.

Consider again the problem of example 1 but replace the amplitude--constraints (13) by the "velocity-constraint"

$$G(u) = \int_0^t u(\tau)\, d\tau \le V , \tag{20}$$

where V - given positive number corresponding to the maximum admissible velocity of the output coordinate.

Since $u \in L^2[0, T]$ and the operator $G(u)$ is acting into the space $C[0, T]$ (i.e. $G : L^2 \to C$), in which the general form of a nonnegative linear functional is described by (3) of sec. 3, the Lagrangean of the present problem becomes

$$\Phi(u, \mu, \lambda) = \frac{1}{2} \int_0^T [u(\tau)]^2 d\tau =$$

$$= \mu_1 \left(\int_0^T (T - \tau) u(\tau) d\tau - x_0 \right) + \mu_2 \int_0^T u(\tau) d\tau +$$

$$+ \int_0^T d\lambda(t) \left[\int_0^t u(\tau) d\tau - V \right] dt,$$

where $\lambda(t)$ is a nondecreasing function with bounded variation, continuous from the right side.

Since

$$\frac{d}{d\gamma} \int_0^T d\lambda(t) \left\{ \int_0^t [u(\tau) + \gamma h(\tau)] d\tau - V \right\} = \int_0^T d\lambda(t) \int_0^t h(\tau) d\tau =$$

$$= \int_0^T h(\tau) \int_\tau^T d\lambda(t),$$

one gets

$$\text{grad}_u \int_0^T d\lambda(t) \left[\int_0^t u(\tau) d\tau - V \right] = \int_\tau^T d\lambda(t),$$

where the function $l(\tau) = \int_\tau^T d\lambda(t) = \lambda(T) - \lambda(\tau)$ is non-increasing.

Then the necessary and sufficient condition for the minimum of the functional (7) subject to the constraints (8), (9), (20) is according to (10) - - (13)

(21) $$\text{grad}_u \Phi(u, \mu, \lambda) = u(\tau) + \mu_1 (T - \tau) + \mu_2 + l(\tau) = 0,$$

(22) $$\int_0^t u(\tau) d\tau \leq V ,$$

$$\int_0^T d\lambda(\tau)\left[\int_0^\tau u(s)\,ds - V\right] = 0, \tag{23}$$

$$d\lambda(\tau) \geq 0 , \tag{24}$$

where $d\lambda(\tau) \geq 0$ denotes the increase of $\lambda(\tau)$ in an arbitrarily small vicinity of τ .

In order to satisfy (21) - (24) , in the general case when (22) is active , we split $[0, T]$ into three subintervals $[0, T_1]$, $[T_1, T_2]$, $[T_2, T]$ and assume

$$\bar{u}(t) = 0, \quad t \in [T_1, T_2]$$

$$l(t) = \text{const}, \quad t \in [0, T_1), \quad t \in (T_2, T]$$

as illustrated by Fig. 5.

Then the condition (21) can be written as

$$\bar{u}(t) + \mu_1(T - t) + \mu_2 + l(T_1) = 0, \quad t \in [0, T_1) , \tag{25}$$

$$\mu_1(T - t) + \mu_2 + l(t) = 0, \quad t \in (T_1, T_2) , \tag{26}$$

$$u(t) + \mu_1(T - t) + \mu_2 = 0, \quad t \in (T_2, T] . \tag{27}$$

Since $l(t)$ should be continuous at T_1, T_2 one obtain from these equations

$$u(t) = \begin{cases} -\mu_1(T_1 - t), & t \in [0, T_1] , \\ 0, & t \in [T_1, T_2] \\ -\mu_1(T_2 - t) - \mu_2, & t \in [T_2, T] \end{cases} \tag{28}$$

$$l(t) = \begin{cases} -\mu_1(T - T_1) - \mu_2, & t \in [0, T_1] , \\ -\mu_1(T - t) - \mu_2, & t \in [T_1, T_2] , \\ 0 , & t \in [T_2, T] \end{cases} \tag{29}$$

In order (9) to be valid it should be

$$T_2 = T - T_1$$

Then from (26) for $t = T_2$ and $l(T_2) = 0$ one gets $\mu_2 = -\mu_1 T_1$.

By (20) and (8) we obtain

$$\int_0^{T_1} \bar{u}(\tau)\, d\tau = \int_0^{T_1} \mu_1 (t - T_1) dt = \frac{\mu_1 T_1^2}{2} = V$$

$$\int_0^{T} \bar{u}(\tau)\, d\tau = \int_0^{T} (T - t)\, \mu_1 (t - T_1)\, dt + \int_{T-T_1}^{T} (T-t)\, \mu_1 (t-T+T_1) dt$$

$$= T_1^2 \mu_1 \left[\frac{T_1}{3} - \frac{T_2}{2}\right] = x_o .$$

From these equations one finds

$$T_1 = \frac{3}{2}\left(T - \frac{x_o}{V}\right), \quad \mu_1 = -\frac{8V}{9\left[T - \frac{x_o}{V}\right]^2}$$

In order to get $T_1 > 0$ it must be $\frac{TV}{x_o} > 1$. In other words the system is optimally controllable, in the sense of sec. 1, if the velocity constraint V is greater then $\frac{x_o}{T}$.

In Fig. 5 a typical plot of $u(t)$, $l(t)$ and the "velocity" $v(t) = \int_0^t \bar{u}(\tau)\, d\tau$ has been shown .

Example 3.

It is required to find such a control $u(t)$ of the system (6), which minimizes the functional

(30) $$F[u] = \int_0^T |u(\tau)|\, d\tau ,$$

subject to the constraints (8), (9), (13).

Since the derivative of the function under the integral sign of (30) is discontinuous we can not use the formula

$$\left.\frac{dF[u + \gamma h]}{d\gamma}\right|_{\gamma = 0}$$

for the derivation of the gradient of $F[u]$. For that reason we represent $u(t)$, as a difference of two nonnegative functions, i.e.

$$u(t) = u_1(t) - u_2(t) \quad ,$$

where

$$u_1(t) = \max[u(t), 0] \; , \quad u_2(t) = \max[-u(t), 0]$$

In the present case the Lagrangean takes the following form

$$\Phi[u, \mu, \lambda] = \int_0^T [u_1(\tau) + u_2(\tau)] d\tau +$$

$$+ \mu_1 \left\{ \int_0^T (T - \tau)[u_1(\tau) - u_2(\tau)] d\tau - x_0 \right\} +$$

(31) $$+ \mu_2 \int_0^T [u_1(\tau) - u_2(\tau)] d\tau + \int_0^T \lambda_1(\tau) [u_1(\tau) - M] d\tau +$$

$$+ \int_0^T \lambda_2(\tau) [u_2(\tau) - M] d\tau$$

where $\lambda_1(\tau), \lambda_2(\tau) \in L^\infty [0, T]$.

Before we write down the conditions of optimality we should derive the differentials of (31)

(32) $$d_{u_1} \Phi[u, \mu, \lambda; h] = \int_0^T [1 + \mu_1(T - \tau) + \mu_2 + \lambda_1(\tau)] \; h(\tau) \, d\tau \; ,$$

(33) $$d_{u_2}\Phi[u,\mu,\lambda;h] = \int_0^T [1-\mu_1(T-\tau)-\mu_2+\lambda_2(\tau)]\,h(\tau)\,d\tau,$$

(34) $$d_{\lambda_1}\Phi[u,\mu,\lambda;h] = \int_0^T [u_1(\tau) - M]\,h(\tau)\,d\tau,$$

(35) $$d_{\lambda_2}\Phi[u,\mu,\lambda,h] = \int_0^T [u_2(\tau)-M]\,h(\tau)\,d\tau.$$

Now we can show that the optimum solution consists of two pulses with the height equal $\pm M$ (see Fig. 6)

(36), (37) $$\bar{u}_1(t) = \begin{cases} M, & 0 \le t \le T_1 \\ 0, & T_1 \le t \le T \end{cases}$$

(38), (39) $$\bar{u}_2(t) = \begin{cases} 0, & 0 \le t \le T-T_1 \\ M, & T-T_1 \le t \le T \end{cases}$$

and

(40) $$\bar{\lambda}_1(t) = \begin{cases} -1-\mu_1(T-t)-\mu_2 \ge 0, & 0 \le t \le T_1 \\ 0, & T_1 \le t \le T \end{cases}$$

(41) $$\bar{\lambda}_2(t) = \begin{cases} 0, & 0 \le t \le T-T_1 \\ -1+\mu_1(T-t)+\mu_2 \ge 0, & T-T_1 \le t \le T \end{cases}$$

The value of T_1 can be determined by setting $\bar{u}(t)$ into (8) which yields

(42) $$\int_0^T (T-t)\,M\,dt - \int_{T-T_1}^T (T-t)M\,dt = MT_1(T-T_1) = x_o.$$

The value of μ_1, μ_2 can be determined using the relations $\bar{\lambda}_1(T_1) = 0$, $\bar{\lambda}_2(T - T_1) = 0$, which yield :

$$-1 - \mu_1(T - T_1) - \mu_2 = 0,$$

$$-1 + \mu_1 T_1 + \mu_2 = 0.$$

Then

$$\mu_1 = \frac{1}{T_1 - \frac{T}{2}}, \quad \mu_2 = \frac{T/2}{\frac{T}{2} - T_1}$$

Using formulae (32) - (41) it is possible to check that

$$d_{u_1} \Phi\left[\bar{u}, \mu, \bar{\lambda}; \bar{u}\right] = d_{u_2} \Phi\left[\bar{u}, \mu, \bar{\lambda}; \bar{u}\right] = 0,$$

$$d_{\lambda_1} \Phi\left[\bar{u}, \mu, \bar{\lambda}; \bar{\lambda}\right] = d_{\lambda_2} \Phi\left[\bar{u}, \mu, \bar{\lambda}; \bar{\lambda}\right] = 0,$$

and

$$d_{u_1} \Phi\left[\bar{u}, \mu, \bar{\lambda}; u\right] \geq 0, \quad d_{u_2} \Phi\left[\bar{u}, \mu, \bar{\lambda}; u\right] \geq 0, \quad u \geq 0,$$

$$d_{\lambda_1} \Phi\left[\bar{u}, \mu, \bar{\lambda}; \lambda\right] \leq 0, \quad d_{\lambda_2} \Phi\left[\bar{u}, \mu, \bar{\lambda}; \lambda\right] \leq 0, \quad \lambda \geq 0,$$

which indicate that the solution (36) - (39) is optimum.

In Fig. 7 the plot of the relation $\frac{T_1}{T}\left[\frac{|x_0|}{MT^2}\right]$ has been shown . It can be observed that the optimum solution exists only if $M \geq \frac{4|x_0|}{T^2}$. In other words, when M is greater than $\frac{4x_0}{T^2}$ the system is optimally controllable in the sense of section 1.

It should be observed that the problems considered in the examples 1 and 3 can be easily extended for the case when the constraints of the control force are given time functions $M_1(t)$, $M_2(t)$, i.e.

$$M_1(t) \leq u(t) \leq M_2(t) .$$

In a similar way the quantity V in the example 2 can be treated as a time function.

It is also possible to consider minimum-time problems, which consist in finding such a control force u(t) which minimizes T subject to the constraints (8), (9) , (10) and

$$\int_0^T |u(\tau)|^p \, d\tau \leq U \, , \quad p \geq 1,$$

where U = given positive number.

If for example $p = 1$ from (36) - (39) one obtains

$$U = \int_0^T |\bar{u}(\tau)| \, d\tau = 2MT_1 \, . \tag{43}$$

By elimination of T_1 from (42), (43) one obtains

$$T = \frac{1}{2}\left[\frac{U}{M} + \frac{\alpha M}{U}\right] , \quad \alpha = 4 \, \frac{|x_0|}{M}$$

In Fig. 8, the plot of that function for $\alpha =$ const has been given. For a given numerical values of α and U/M one can read from that plot a correspondig minimum time value $T = \bar{T}$, which determines the optimum solution. Then using (42) the corresponding value of T_1 can be derived.

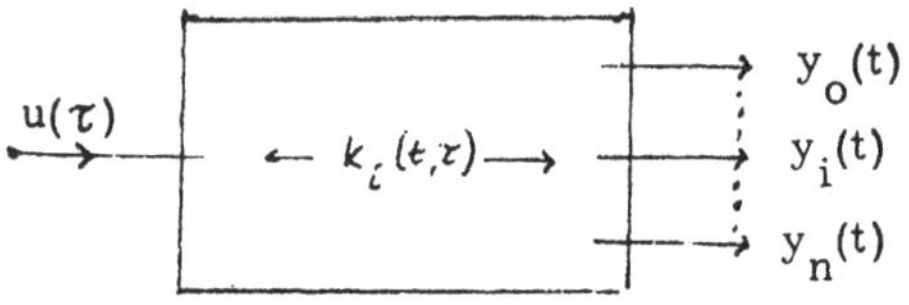

Fig. 1

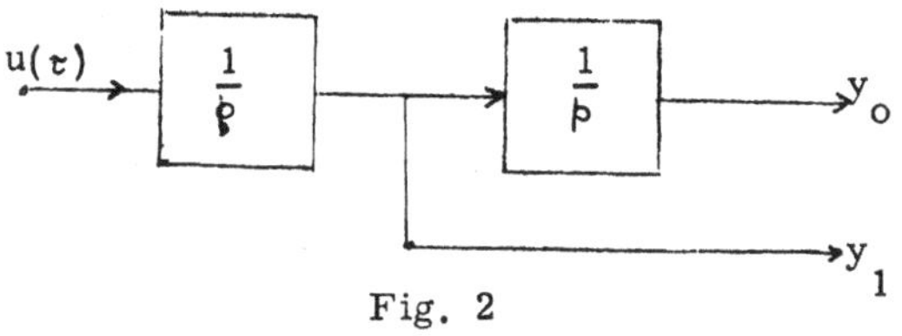

Fig. 2

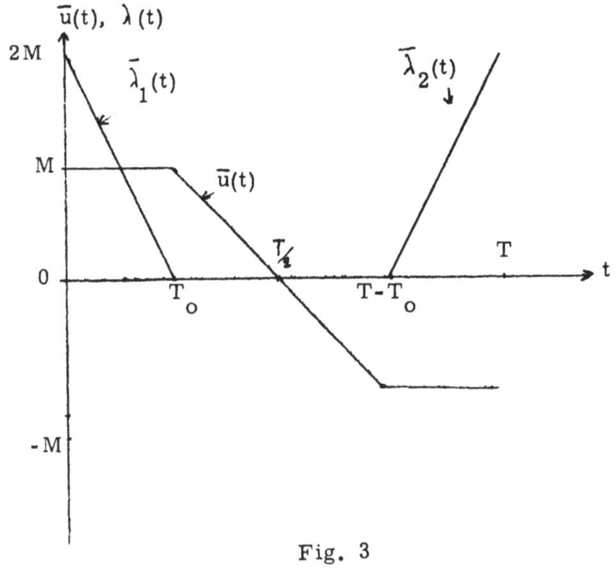

Fig. 3

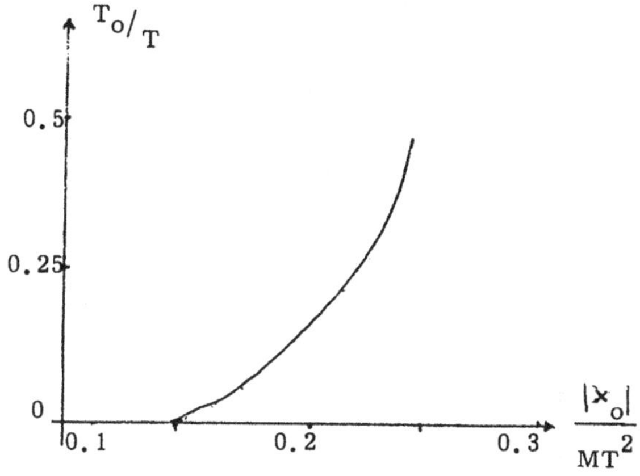

Fig. 4

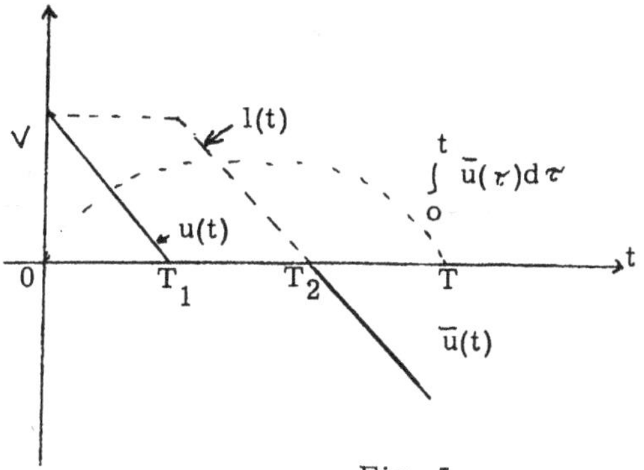

Fig. 5

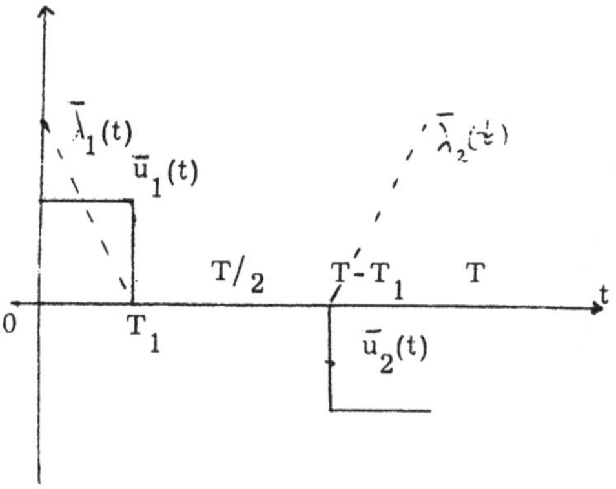

Fig. 6

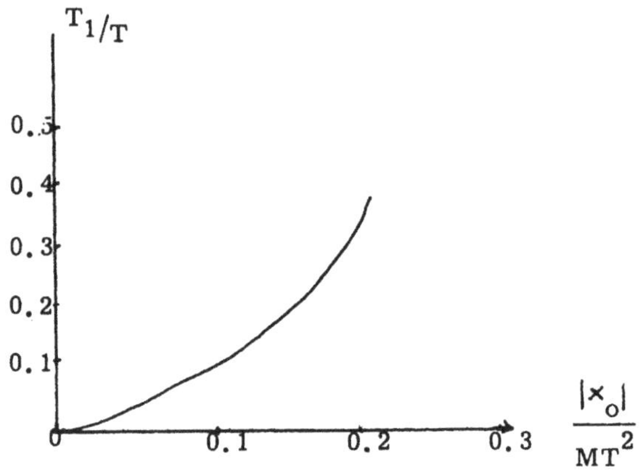

T_1/T
0.5
0.4
0.3
0.2
0.1
0
0.1
0.2
0.3
$\frac{|x_o|}{MT^2}$

Fig. 7

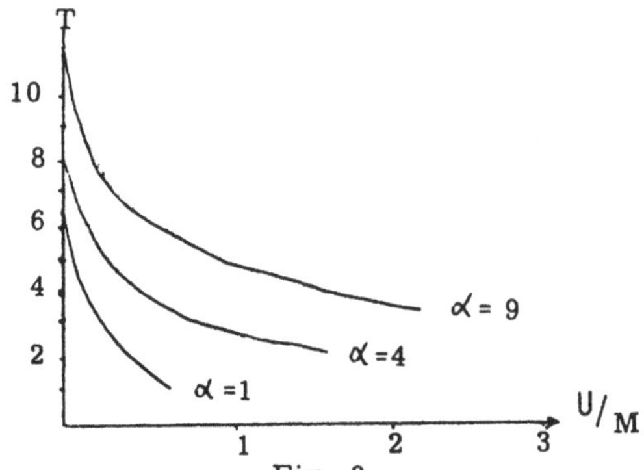

T
10
8
6
4
2
$\alpha = 9$
$\alpha = 4$
$\alpha = 1$
1
2
3
U/M

Fig. 8

R. Kulikowski

References

1. Hurwicz L. : Programming in linear spaces, in K.I. Arrow, L. Hurwicz, H. Uzawa: Studies in linear and nonlinear programming, Stanford 1958.

2. Kalman R.E. : On the general theory of control systems. Proc. IFAC Congr. Moscow 1960 vol. 1 pp. 481-493. London 1961.

3. Kalman R.E. , Ho Y.C., Narendra K.S. : Controllability of linear dynamical systems, in Contribution to differential equations. Vol. 1. New York 1962.

4. Kulikowski R. : On optimal control with integral and magnitude type of constraints. Warszawa 1967 Prace Instytutu Automatyki PAN z. 67.

5. Kulikowski R. : On optimum control of nonlinear, dynamic industrial processes. Archiwum Automatyki i Telemechaniki 1967 z. 1.

6. Lusternik L.A. , Sobolev W.I. : Elements of functional analysis. Moscow 1951 (in Russian, English translation available).

7. Majerczyk-Gómułka J. , Makowski K: Wyznaczanie optymalnego sterowania procesami dynamicznymi metodą funkcjonałow Lagrange' a. Archiwum Automatyki i Telemechaniki 1968 z. 2, 3.

8. Pontryagin L.S., Boltianskii V.G. , Gamkrelidze R.V. , Mishchenko E.F. : The mathematical theory of optimal processes. New York 1962. English translation by K.N. Trirogoff.

9. Weinberg M.M. : Variational methods of investigation of nonlinear operators. Moscow 1956 (in Russsian, English translation available) .

CENTRO INTERNAZIONALE MATEMATICO ESTIVO

(C.I.M.E.)

A. STRASZAK

SUPERVISORY CONTROLLABILITY

Corso tenuto a Sasso Marconi dal 1 al 9 luglio 1968

SUPERVISORY CONTROLLABILITY

by

A. Straszak

(Institute for Automatic Control - Polish Academy of Sciences, Warsaw-)

1. Introduction

Multilevel control systems are of increasing importance in the application of complex automatic control in industrial and non-industrial field. The multilevel approach to the synthesis or design of the multivariable control system is the only one acceptable for real large-scale systems. A multilevel structure arises rather naturally in practice. However, the theoretical foundation of this concept is still on the early stage of the research, since conventional multivariable optimization and control theories can not be used directly. [1] [4] [5] [6]

II. The Problem

Let us consider a set of optimal control systems.

$$\frac{d x_i}{dt} = f_i(x_i, u_i)$$

$$u_i = c_i(x_i) \qquad i = 1, 2, \ldots, k.$$

where

$$x_i = (x_{oi}, x_{1i}, \ldots, x_{n_i i})$$

$$u_i = (u_{1i}, u_{2i}, \ldots, u_{m_i i})$$

$$x_{oi} = \int_{t_{o_i}}^{t_{1_i}} f_{oi}(x_i, u_i)\, dt \text{ - index}$$

of performance of the control system

$$u_i \in U_i$$

Each of the optimal control systems is optimal due to some fixed resources (for example: fuel, energy) or another factor (for example: a price of the control given to the system).

Suppose that the set of the optimal control systems have common sources of this resources or resource - like factors.

Now, we may formulate the following problem.

Firstly, can we improve the optimality of the optimal control system which belongs to the given set of the optimal control systems by using the multilevel control procedure.

Secondly, can we improve the overall optimality (in the sense of the global index of performance) of the set of the optimal control system by using the multilevel control procedure.

However, we can not answer these questions before the study of controllability (coordinability) of the optimal control system. Therefore, first of all we must know if it is possible to control the optimal control system and how to do it.

III Supervisory Controllability

The optimal control law

$$u_i = c_i(x_i)$$

is optimal only for the given dynamic process

$$\frac{dx_i}{dt} = f_i(x_i, u_i)$$

and given constraints

$$u_i < r_i^1$$

and/or

$$\int_0^{T_i} (u_i(\tau))^2 d\tau < r_i^2$$

and/or

$$\int_0^{T_i} |u_i(\tau)| d\tau < r_i^3$$

and so on.

Therefore the dynamic process f_i with the controller

$$u_i = c_i(x_i, f_i, r_i^j)$$

must be controllable in the working domain of the state space $D_i \subseteq X$ due to the given resources r_i^j.

Now, we can formulate the supervisory controllability:

We say that the dynamic process f_i is a supervisory controllable if for any given number $\eta > 0$, the dynamic process f_i is controllable in the working domain D_i or working subdomain $D_i^* \in D_i$ due to given resources $\eta\, r_i^j$ and the interval $< x_{oi}(\eta) >$ consists of the more than one point for different η.

Example.

Let us consider the dynamic process

$$\frac{dx}{dt} = \begin{bmatrix} 0 & 0 \\ 1 & 0 \end{bmatrix} x + \begin{bmatrix} 1 \\ 0 \end{bmatrix} u$$

with the working domain $< x_1(t=0) = 0,\ x_2(t=0) = 0,\ x_1(t=T) = 0,\ x_2(t=T) = x_2^* >$

and

$$x_o = \int_o^T dt = T \text{ - minimum time index of performance.}$$

Using the functional analysis methods for this optimal control problem [3], we obtain that

$$T = \sqrt{\frac{|x_2^*|}{\text{const.}\,\eta}} \qquad \text{for} \quad |u| \leq r = 1,$$

$$T = \frac{1}{4}\sqrt{\frac{|x_2^*|}{\text{const.}\,\eta}} \qquad \text{for} \quad \int_0^T |u(\tau)|\, d\tau \leq r = 1,$$

$$T = 2\sqrt[3]{\frac{|x_2^*|\, k_1}{k_2\sqrt{\eta}}} \qquad \text{for} \quad \int_0^T \left[u(\tau)\right]^2 d\tau \leq r = 1 .$$

Therefore this dynamic process with $f_o = 1$ has the supervisory controllability.

4. Supervisory Control.

Assume that we have a set of the supervisory controllable optimal control systems, therefore we have relations

$$x_{oi} = g_{oi}(x_i, r_i^j)$$

where

$$x_{oi} = \min_{u_i} x_{oi}$$

and vector index of performance

$$x_{og} = (\hat{x}_{o1}, \hat{x}_{o2}, \ldots, \hat{x}_{ok})$$

By introducing the global index of performance

$$I_g = \max_i \hat{x}_{oi}$$

or

$$I_g = \sum_{i=1}^{k} \hat{x}_{oi}$$

we can formulate the optimization problem for a supervisory controller:

Minimize

$$I_g = \max_i \hat{x}_{oi}$$

subject to

$$r_1^j + r_2^j + \ldots + r_k^j \leqslant R^j$$

or :

Minimize

$$I_g = \sum_{i=1}^{k} \hat{x}_{oi}$$

subject to

$$r_1^j + r_2^j + \ldots + r_k^j \leq R^j$$

which can be solved by the supervisory controller by using the mathematicla programming machinery 2 , 7 .

References

1. Mesarovic, M. D. : Advances in multilevel control Proc. IFAC Tokyo Symposium on system Engineering, Tokyo 1965.
2. Karlin, S: Mathematical Methods and Theory in Games, Programming and Economics. Pergamon Press. London 1959.
3. Kulikowski, R. : Optimal Control as a Function of Plant Parameters. Arch. Aut. i Tel. n. 2 1961 . Warsaw (Polish).
4. Kulikowski, R. : Optimum control of aggregated multilevel system, Proc. III IFAC Congress , 1966. London 1966.
5. Straszak, A. : On the structure synthesis problem in multilevel control systems. Proc. IFAC Tokyo Symposium on System Engineering . Tokyo 1965.
6. Straszak, A. : Multilayer and Multilevel control Structures. Proc. on Neural Network Springer - Verlag. Heidelberg 1968.
7. Straszak A. : Suboptimal supervisory control. Functional analysis and optimization New York. Academic Press 1966.

CENTRO INTERNAZIONALE MATEMATICO ESTIVO

(C.I.M.E.)

LECTURES ON CONTROLLABILITY AND OBSERVABILITY

L. WEISS

(University of Maryland)

Corso tenuto a Sasso Marconi (Bologna) dal 1 al 9 Luglio 1968

CONTENTS

L. Weiss

1. INTRODUCTION

These lecture notes are devoted to a detailed examination of the fundamental system-theoretic concepts of Controllability, Observability, Reachability and Determinability, and of the roles they play in certain specific areas of research in modern system theory. The mathematical models employed in this study range in sophistication from linear, constant coefficient, differential equations to nonlinear, time-varying functional-differential equations. Much of the material, especially that dealing with functional-differential systems is of very recent origin, and in many cases enables well known older results in controllability theory to be embedded within newer more general ones. An overall objective has been to incorporate a certain amount of self-containment and breadth in the presentation. Nonetheless, various aspects of the subject have not been covered (it might be well to point out the obvious fact that there are as many types of problems one could consider as there are different types of system models). Some notable omissions which a prospective lecturer on the subject may wish to fill in are: (1) Controllability for partial differential equations and differential equations on Banach spaces. (2) The relationship of controllability to the problem of system stabilization.

A series of seven lectures bases on the material in these notes presented by the author at the Centro Internazionale Matematico

Estivo at Pontecchio Marconi, Italy, in July, 1968. I am indebted to Professors E. Bompiani, R. Conti, and G. Evangelisti for the privilege of participating in the C.I.M.E. course on Controllability and Observability.

2. EXISTENCE AND UNIQUENESS OF SOLUTIONS TO DELAY-DIFFERENTIAL EQUATIONS

Although our concern is with a restricted class of delay-differential equations, it is instructive to present an existence theorem for a more general class of equations. The results, whose proofs are not reproduced here, are due to Driver [10]. Notation is as follows.

D = open connected set in R^n

If $x = \text{col}(x_1,\dots,x_n) \in R^n$, then $||x|| = \max\limits_i |x_i|$

If $[a,b]$ is an interval in R and $\phi : [a,b] \to R^n$, then

$||\phi||_{[a,b]} = \sup\limits_{a \leq t \leq b} ||\phi(t)||$.

$C([a,b],K)$ = class of continuous functions mapping $[a,b]$ into $K \subset R^n$.

Consider the delay-differential system

$$\frac{dx}{dt} = f(t,x(\cdot)) \ , \ t \in (t_o,\gamma) \ , \ \gamma \leq \infty \tag{2.1}$$

where $f(t,\phi(\cdot)) \in R^n$ for $t \in (t_o,\gamma)$, $\phi \in C([\alpha,t],D)$ for some $\alpha \in (-\infty,t_o]$. Then

<u>Definition 2.1</u>. (i) f is <u>continuous</u> in t if $f(t,\phi(\cdot))$ is a continuous function of t for $t_o \leq t \leq \gamma$ for any $\phi \in C([\alpha,\gamma],D)$.

(ii) f is <u>locally lipschitzian</u> in ϕ if for every $\beta \in [t_o,\gamma)$ and every compact set $S \subset D$ there exists a constant $K_{\beta,uc}$ such that

$$||f(t,\phi_1(\cdot)) - f(t,\phi_2(\cdot))||_{[\alpha,t]} \leq K_{\beta,uc}||\phi_1 - \phi_2||_{[\alpha,t]}$$

for all $t \in [t_o,\beta)$, all $\phi_1, \phi_2 \in C([\alpha,t],S)$.

Definition 2.2. (i) Given an initial function $\phi(\cdot) \in C([\alpha,t_o],D)$, a solution to (1) is a function $x(\cdot) \in C([\alpha,\beta),D)$ $t_o < \beta \leq \gamma$, such that $x(t) = \phi(t)$ for all $t \in [\alpha,t_o]$ and $x(t)$ satisfies (1) on (t_o,β) .

(ii) The solution at time t generated from initial time t_o and initial function ϕ is denoted by $x(t;t_o,\phi)$. This solution is unique if any other solution $y(t;t_o,\phi)$ is identical to it as far as both are defined.

Theorem 2.2. Consider the system (2.1) and let $f(t,\psi(\cdot))$ be continuous in t and locally lipschitzian in ψ . Let $\phi \in C([\alpha,t_o],D)$. Then there exists a unique solution $x(t) = x(t;t_o,\phi)$ on $[\alpha,\beta)$, $t_o < \beta \leq \gamma$, and if $\beta < \gamma$ and β cannot be increased, then for any compact set $G \subset D$ there exists a sequence of real numbers $t_o < t_1 < t_2 < .. \to \beta$ such that $x(t_k) \in D - G$, $k = 1,2,...,$ i.e., as $t \to \beta$, $x(t)$ comes arbitrarily close to D or is unbounded.

Corollary 2.3. Let $D = R^n$ and let $f(t,\psi(\cdot))$ be continuous in t and ψ , and linear in ψ . Then for every $\phi \in C([\alpha,t_o],R^n)$ there exists a unique solution $x(t) = x(t;t_o,\phi)$ on the entire interval $[\alpha,\gamma)$.

3. REPRESENTATION OF SOLUTIONS FOR LINEAR DELAY-DIFFERENTIAL SYSTEMS

In this section, which is based heavily on the work of Hale and Meyer [12], we consider the equations of the form

$$\frac{dx}{dt} = f(t,x(\cdot)) + u(t) \tag{3.1}$$

with initial function space $C([t_o - h,t_o],R^n) = \mathcal{B}$ where $f(t,x(\cdot))$ is linear in $x(\cdot)$ and depends only on values of $x(s)$ for $t - h \leqslant s \leqslant t$. It is further assumed that $||f(t,\phi(\cdot))|| \leqslant L(t)||\phi||_{[t-h,t]}$ for all $\phi \in \mathcal{B}$, all t, where $L(\cdot)$ is continuous and positive. The control function $u(\cdot)$ belongs to the class of functions which are measurable and bounded on every finite time interval. Our objective is to outline the derivation of a "variation of parameters" formula. We note first that (3.1) is equivalent to the functional integral equation

$$\begin{cases} x(t) = \phi(t) \ , \ t \in [t_o - h,t_o] \\ x(t) = \displaystyle\int_{t_o}^{t} f(s,x(s))ds + \int_{t_o}^{t} u(s)ds + \phi(0) \ , \ t \geqslant t_o \ . \end{cases} \tag{3.2}$$

to which a unique solution exists by Theorem 2.2. The hypotheses on f allow application of the Riesz Representation Theorem to establish existence of an $n \times n$ matrix valued function η defined on $(-\infty,\infty) \times [-h,0]$ such that $f(t,\psi(\cdot)) = \int_{-h}^{0} [d_\tau \eta(t,\tau)]\psi(\tau)$ for all $\psi \in \mathcal{B}$. Moreover, $\eta(t,\cdot)$ is of bounded variation on $[-h,0]$ for each t. Now let $L_1([t_o,\tau),R^n)$ denote the space of functions with range in R^n which are Lebesgue integrable over $[t_o,\tau)$. Then we have

<u>Theorem 3.1.</u> Let $x(\cdot,t_o,\phi,u)$ be the solution of (3.1) (or (3.2)) with control $u \in L_1([t_o,\tau),R^n)$ for all $\tau \geq t_o$ and with $\phi \in \mathcal{B}$. Then

$$x(t;t_o,\phi,u) = x(t;t_o,\phi,0) + \int_{t_o}^{t} K(t,s)u(s)ds \ , \ t \geq t_o \tag{3.3}$$

where $K(t,s)$ is defined for $s \leq t - h$, $K(t,\cdot) \in L_\infty((t_o,t],R^{n^2})$ for each t, and $K(t,s) = \frac{\partial W(t,s)}{\partial s}$ almost everywhere, where $W(t,s)$ is the unique solution of the equations

$$\begin{cases} W(t,s) = 0 \quad \text{for all} \quad t \in [s-h,s] \\ W(t,s) = \int_s^t \int_{-h}^{0} \{d_\tau \eta(\xi,\tau)\} \, W(\tau+\xi,\xi)d\xi - (t-s)I \ , \ s \leq t \ . \end{cases} \tag{3.4}$$

<u>Proof:</u> Let $u(\cdot) \in L_1([t_o,t],R^n)$. Then $M(u) \triangleq x(t;t_o,0,u)$ is a continuous linear operator mapping $L_1([t_o,t],R^n)$ into R^n and by the Riesz Representation Theorem, there exists an $n \times n$ matrix $\mathfrak{X}(t;t_o,\cdot) \in L_\infty([t_o,t],R^{n^2})$ uefined for all $t \geq t_o$ such that

$$x(t;t_o,0,u) = \int_{t_o}^{t} X(t;t_o,\tau)u(\tau)d\tau \ .$$

It is easily shown [12] that X is independent of t_o. Hence let $K(t,\tau) = X(t;t_o,\tau)$, $t \varepsilon (-\infty,\infty)$, $\tau \varepsilon (-\infty,t]$ and let $K(t,\tau) = 0$ for $\tau \varepsilon (t,t+h]$. For any $\eta \varepsilon (-\infty,\infty)$, let $W(t,\eta) = -\int_{\eta}^{t} K(t,\tau)d\tau$ for $t \geqslant \eta$ and $W(t,\eta) = 0$ for $t \varepsilon [\eta - h,\eta]$. Then W satisfies (3.4) and K is as stated in the theorem.

The linear system to be discussed in some detail from a controllability standpoint is of the form

$$\frac{dx}{dt} = A(t)x(t) + B(t)x(t-h) + C(t)u(t) \tag{3.5}$$

where $x(t) \varepsilon R^n$, $u(t) \varepsilon R^p$, and $A(\cdot)$, $B(\cdot)$, $C(\cdot)$ are continuous functions. The solution of (3.5) can be represented as in (3.3) and it is easily checked that the function $K(t,s)$ satisfies the partial differential equations [2]

$$\begin{cases} \dfrac{\partial K(t,s)}{\partial s} = -K(t,s)A(s) - K(t,\, s+h)B(s+h) \ , \quad t_o \leqslant s < t-h \\[2ex] \dfrac{\partial K(t,s)}{\partial s} = -K(t,s)A(s) \ , \quad t-h \leqslant s \leqslant t \\[2ex] K(t,t) = I \ . \end{cases} \tag{3.6}$$

4. DEFINITIONS OF CONTROLLABILITY

Consider the nonlinear delay-differential system

$$\frac{dx}{dt} = f(t,x(t),x(t-h),u(t)) \ , \ t \geq t_o \tag{4.1}$$

where $x(t) \in R^n$, $u(t) \in R^p$, and u is measurable and bounded on every finite time interval (such controls will be called "admissible"). The delay is represented by a real scalar $h > 0$ and it is assumed that $f \in C^1$ in all its arguments and $f(t,0,0,0) = 0$ for all t . The initial function space is the space $\mathcal{B}$ as defined earlier.

Definition 4.1. The system (4.1) is R^n - controllable if for any $\phi \in \mathcal{B}$ there exists $t_1 = t_1(\phi) \in (t_o,\infty)$ and an admissible control segment $u_{[t_o,t_1]}$ such that $x(t_1;t_o,\phi,u) = 0$. If t_1 is independent of ϕ , we speak of fixed-time R^n - controllability, and if $t_1 - t_o$ can be made arbitrarily small we speak of differential R^n - controllability.

While this definition turns out to be quite useful, it does not reflect the fact that the state space of (4.1) is a function space and that one can conceive of control problems in which the state of the system is to be transferred to a point (or region) in function space. Hence it makes sense to also consider the following definition.

Definition 4.2. The system (4.1) is controllable to the origin with respect to the space of initial functions $\mathcal{B}$ if for any $\phi \in \mathcal{B}$ there exists $t_1 = t_1(\phi) \in (t_o, \infty)$ and an admissible control segment $u_{[t_o, t_1+h]}$ such that $x(t;t_o,\phi,u) = 0$ for all $t \in [t_1, t_1 + h]$. Although controllability to the origin does not imply controllability to some other point in function space, it is possible to obtain results for the latter problem using an approach similar to that presented in the sequel (see Weiss [23]).

5. CONTROLLABILITY OF LINEAR DELAY-DIFFERENTIAL SYSTEMS

We begin with the following Lemma.

Lemma 5.1. The system (3.5) is R^n - controllable if there exists $t_1 > t_o$ such that

$$\text{(5.1)} \qquad \operatorname{rank} \int_{t_o}^{t_1} K(t_1,\eta)C(\eta)C'(\eta)K'(t_1,\eta)d\eta = n$$

where the prime (') indicates transpose.

Proof: Let $\mathcal{C}$ be the matrix in (5.1) whose rank is n. The Lemma follows by substituting

$$u(s) = -C'(s)K'(t_1,s)\mathcal{C}^{-1}x(t_1;t_o,\phi,0)$$

in (3.3), for then $x(t_1;t_o,\phi,u) = 0$. The question of the necessity of (5.1) involves the concept defined below.

Definition 5.1. A system (3.5) is pointwise complete if for each t there exists a set of initial functions $\phi_i^t \varepsilon \mathcal{B}$, $i = 1,\ldots, n$, such that the set $x(t;t_o,\phi_i^t,0)$, $i = 1,\ldots, n$ forms a basis for R^n .

It is easy to construct an example to show that not all time-varying systems (3.5) are pointwise complete. We conjecture, however, that all constant coefficient systems of the form (3.5) are pointwise complete.

Lemma 5.2. If the system (3.5) is pointwise complete, then (5.1) is necessary and sufficient for fixed-time R^n - controllability.

Proof: For any $\phi \varepsilon \mathcal{B}$, suppose there exists a fixed time $t_1 > t_o$ and an admissible control $u_{[t_o,t_1]}$ such that $x(t_1;t_o,\phi,u) = 0$ but (5.1) does not hold. Then there exists a nonzero vector $x_1 \varepsilon R^n$ such that $x_1'K(t_1,s)C(s) = 0$ for all $s \varepsilon [t_o,t_1]$. Then $x_1'x(t_1;t_o,\phi,0) = 0$. By hypothesis, ϕ can be chosen so that $x(t_1;t_o,\phi,0) = x_1$. Then $x_1'x_1 = 0$ which is a contradiction.

Theorem 5.3. A system (3.5) is controllable to the origin with respect to $\mathcal{B}$ if and only if

(i) it is R^n - controllable

(ii) for each $\phi \varepsilon \mathcal{B}$ and for some corresponding t_1 and admissible $u_{[t_o,t_1]}$ such that $x(t_1;t_o,\phi,u) = 0$, the equation

(5.2) $$C(t)u(t) = - B(t)x(t - h;t_o,\phi,u)$$

has an admissible solution $u^*(\cdot)$ defined on $(t_1,t_1 + h)$.

Proof: The necessity of (i) is obvious. Now, given $\phi \in \mathcal{B}$, let $u_{[t_o,t_1]}$ be such that $x(t_1;t_o,\phi,u) = 0$. If (5.2) holds, then on the interval $(t_1,t_1 + h)$, the system (3.5) becomes $\dot{x}(t) = A(t)x(t)$, $x(t_1) = 0$. It then follows that $x(t) = 0$ for all $t \in [t_1,t_1 + h]$.

Conversely, if (3.5) is controllable to the origin with respect to $\mathcal{B}$, then for each $\phi \in \mathcal{B}$ there exists some $t_1 > t_o$ and control $u_{[t_o,t_1+h)}$ such that $x(t;t_o,\phi,u) = 0$ for all $t \in [t_1,t_1 + h]$. This implies (i) and the uniqueness theorem for delay equations implies (ii).

Remark: The major element in the controllability problem for 3.5 is the solution to (5.2). Clearly, an admissible solution will exist on $(t_1,t_1 + h)$ if and only if the right side of (5.2) is in the range of $C(t)$ almost everywhere on the interval. A sufficient condition for the latter to hold is the existence of an $n \times p$ matrix $D(t)$ with bounded measurable elements such that $B(t) = C(t)D(t)$ almost everywhere on $(t_1,t_1 + h)$.

6. LOCAL R^n - CONTROLLABILITY OF NONLINEAR DELAY-DIFFERENTIAL SYSTEMS

In this section, we generalize the results of Lee and Markus [19] to the case of delay-equations.

Definition 6.1. The system (4.1) is locally R^n - controllable to the origin with respect to $\mathcal{B}$ if it is R^n - controllable to the origin with respect to a neighborhood $N(0^{\mathcal{B}})$ where

$$(6.1)\quad \begin{cases} A(t) = \frac{\partial f}{\partial x}(t,0,0,0) \\ B(t) = \frac{\partial f}{\partial x_d}(t,0,0,0)\ , \text{ where } x_d(t) = x(t-h) \\ C(t) = \frac{\partial f}{\partial u}(t,0,0,0)\ . \end{cases}$$

Theorem 6.1. A system (4.1) is locally R^n - controllable to the origin with respect to $\mathcal{B}$ if its first variation about the zero-solution satisfies the condition that

there exists $t_1 > t_o$ such that (5.1) holds

Proof: We introduce a parameter ξ into the control u and define

$$(6.2)\quad u^{\xi}(t) \equiv u(t,\xi) = C'(t)K'(t_1,t)\xi \quad \text{for} \quad t_o \leq t \leq t_1\ .$$

It should be noted that $u(t,0) = u^{\circ}(t) = 0$ for $t \in [t_o,t_1]$, and that if $\phi \equiv 0$ then $x(t;t_o,0^{\mathcal{B}},u^{\circ}) = 0$ on $[t_o - h,t_1]$.

Define the Jacobian matrix $J(t)$ by

$$J(t) = \left.\frac{\partial x(t;t_o,0^{\mathcal{B}},u^{\xi})}{\partial \xi}\right|_{\xi=0} . \tag{6.3}$$

Since $\phi \equiv 0$, the solution of (4.1) is written as

$$x(t;t_o,0^{\mathcal{B}},u^{\xi}) \equiv x(t,\xi) = \int_{t_o}^{t} f(\tau,x(\tau),x_d(\tau),u^{\xi}(\tau)d\tau \ , \ t_o \leqslant t \leqslant t_1 + h \ .$$

Then we have

$$J(t) = \left.\frac{\partial x}{\partial \xi}\right|_{\xi=0} = \int_{t_o}^{t} [A(\tau)J(\tau) + B(\tau)J(t-h) + C(\tau)\frac{\partial u}{\partial \xi}(\tau,0)]d\tau$$

where A , B , C are as given in (6.1). Differentiating we obtain

$$\dot{J}(t) = A(t)J(t) + B(t)J(t-h) + C(t)\frac{\partial u}{\partial \xi}(t,0) \ , \ t_o \leqslant t \leqslant t_1 + h \ . \tag{6.4}$$

But, from (6.2)

$$\frac{\partial u}{\partial \xi}(t,0) = C'(t)K'(t_1,t) \ , \quad t_o \leqslant t \leqslant t_1$$

and

$$C(t)\frac{\partial u}{\partial \xi}(t,0) = -B(t)J(t-h) \ , \ t_1 < t < t_1 + h \ .$$

Therefore,

$$(6.5)\quad \dot{J}(t) = A(t)J(t) + B(t)J(t-h) + C(t)C'(t)K'(t_1,t)\ ,\quad t_o \leq t \leq t_1\ .$$

The solution of (6.5) over the interval $[t_o,t_1]$ is then

$$(6.6)\quad J(t) = \int_{t_o}^{t} K(t,s)C(s)C'(s)K'(t_1,s)\,ds \qquad t_o \leq t \leq t_1\ .$$

By hypothesis, $\det J(t_1) \neq 0$, and so, application of the implicit function theorem [8] yields a unique continuous map $\pi : N(0^{\mathcal{B}}) \to R^n$ such that if $\phi \in N(0^{\mathcal{B}})$ then the equation $x(t_1;t_o,\phi,\xi) = \dot{0}$ has an admissible solution $\xi = \pi(\phi)$. This proves Theorem 6.1.

It is worth remarking that the hypothesis of Theorem 6.1 is necessary as well as sufficient in order to yield a nonsingular Jacobian matrix (6.3) at the time t_1 . To see this, we need merely note that regardless of the choice of $u(t,\xi)$, the function $J(t_1)$ would have the form

$$J(t_1) = \int_{t_o}^{t_1} K(t_1,s)C(s)Z(s)\,ds$$

where $Z(\cdot) \in L_2[t_o,t_1]$. The necessity of the hypothesis then follows from the Lemma below, which is due to Hermes [13] .

<u>Lemma 6.2.</u> Let $H(\cdot)$ be an $n \times p$ matrix of functions in $L_2[t_o,t_1]$ for any finite $t_1 > t_o$. Then, a necessary and sufficient condition that there exist a $p \times n$ matrix, $V(t)$, of functions in $L_2[t_o,t_1]$ such that $A = \int_{t_o}^{t_1} H(\eta)V(\eta)\,d\eta$ is nonsingular is that $\mathcal{D} = \int_{t_o}^{t_1} H(\eta)H'(\eta)\,d\eta$ is nonsingular.

Proof. Sufficiency is trivial (let $V(t) = H'(t)$. For necessity, assume there exists $V(t)$ such that $\mathcal{A}$ is nonsingular, but $\mathcal{D}$ is singular. Then there exists a constant vector $c \neq 0$ such that $c \mathcal{D} c' = 0$ and since $H(t)H'(t)$ is positive semidefinite we have $cH(t) = 0$ almost everywhere on $[t_o, t_1]$. But then $c \mathcal{A} = 0$ which contradicts the nonsingularity of $\mathcal{A}$.

7. CONTROLLABILITY AND OBSERVABILITY FOR ORDINARY LINEAR AND NONLINEAR DIFFERENTIAL SYSTEMS

Much of what has already been presented for delay equations is directly applicable to the controllability problem for ordinary differential equations. The special nature of the latter, however, allows somewhat sharper results to be obtained. Our emphasis is on linear equations although the nonlinear problem is also discussed in the sequel. Consider the system

$$(7.1) \qquad \frac{dx}{dt} = A(t)x(t) + C(t)u(t)$$

$$y(t) = H(t)x(t)$$

where $x(t) \in R^n$, $u(t) \in R^p$, $y(t) \in R^r$, and $A(\cdot)$, $C(\cdot)$, $H(\cdot)$ are continuous functions of time. We shall find it convenient to refer to the adjoint system to (7.1) defined as the system

$$(7.2) \qquad \frac{dz}{dt} = -A'(t)z(t) + H'(t)\tilde{u}(t)$$

$$\tilde{y}(t) = C'(t)z(t) .$$

It is easily shown that if $\Phi(t,\tau)$ is the transition matrix (matrizant) for (7.1), i.e., if $\Phi(t,\tau)$ satisfies

$$\frac{d\Phi}{dt} = A(t)\Phi(t,\tau) \tag{7.3}$$

$$\Phi(t,t) = I$$

then $\Phi'(\tau,t)$ is the transition matrix for (7.2).

We present the following definitions.

<u>Definition 7.1.</u> A state x_o of the system (7.1) is controllable from time τ if there exists $t > \tau$, t finite, and an admissible input segment $u_{[\tau,t]}$ such that the phase (τ,x_o) is transferred to the phase $(t,0)$. Otherwise x_o is <u>uncontrollable from</u> τ. If every (no) state is controllable from τ, the <u>system is controllable (uncontrollable) from</u> τ. Controllability (uncontrollability) of the system from all τ is denoted by <u>complete controllability (uncontrollability)</u>. If $t - \tau$ in these definitions can be made arbitrarily small we speak of <u>differential controllability</u> from τ.

<u>Definition 7.2.</u> A state x_o of the system (7.1) is <u>reachable at</u> time τ (older terminology: <u>anticausal controllable</u>) if there exists a finite value of time $t < \tau$, and an admissible control segment $u_{[t,\tau]}$ such that the phase $(t,0)$ is transferred to the phase (τ,x_o). Otherwise x_o is <u>unreachable at</u> τ. If every (no) state is reachable at τ, the <u>system is reachable (unreachable) at</u> τ. Reachability (unreachability) of the system at all τ is denoted by <u>complete</u>

reachability (unreachability). If $|t - \tau|$ can be made arbitrarily small, we speak of differential reachability at τ.

Definition 7.3. A state x_o of the system (7.1) is observable from time τ if with respect to the adjoint system that state is controllable from time τ. Otherwise the state is unobservable from τ. Remaining definitions of system observability, unobservability, complete observability, and differential observability follow analogously from Def. 7.1 above.

Definition 7.4. A state x_o is determinable at time τ if, with respect to the adjoint system, that state is reachable at time τ. Otherwise the state is undeterminable at τ. Remaining definitions of system determinability, undeterminability complete determinability and differential determinability follow analogously from Def. 7.2 above.

Remark: In earlier papers of the writer [20], [21], the concepts of "reachability" and "observability" were denoted respectively by "anticausal controllability" and "anticausal observability" while the present difinition of "determinability" corresponds to the older definition of "observability". The rationale behind the terminology will soon become apparent.

Now define the "Controllability" and "Determinability" matrices by the relations

$$(7.4) \qquad \mathcal{C}(t,\sigma) = \int_t^\sigma \Phi(t,\eta)C(\eta)C'(\eta)\Phi'(t,\eta)d\eta$$

$$(7.5) \qquad \mathcal{D}(t,\sigma) = \int_t^\sigma \Phi'(\eta,t)H'(\eta)H(\eta)\Phi(\eta,t)d\eta \ .$$

Let $\mathcal{R}[\cdot]$ denote the range of $[\cdot]$. Then we have

Theorem 7.1. $\mathcal{R}[C(t,\sigma)]$, $\sigma \geq t$ is monotone nondecreasing with increasing σ .

Proof: $C(t,\sigma)$ is a Gramian matrix and has the property that for any $x \in R^n$, $0 \leq x'C(t,\sigma_1)x \leq x'C(t_1\sigma_2)x$ for $\sigma_1 < \sigma_2$. Hence $x \in \mathcal{K}[C(t,\sigma_2)] \Longrightarrow x \in \mathcal{K}[C(t,\sigma_1)]$ where $\mathcal{K}[\cdot]$ denotes Kernel $[\cdot]$. Therefore $\mathcal{K}[C(t,\sigma_1)] \geq \mathcal{K}[C(t,\sigma_2)]$ and by orthogonal complementation $\mathcal{R}[C(t,\sigma_1)] \leq \mathcal{R}[C(t,\sigma_2)]$.

Corollary 7.2. There exists a positive C^1 function $\mu(t)$ such that

$$\bigcup_{\sigma>t} \mathcal{R}[C(t,\sigma)] = \mathcal{R}[C(t,t+\mu(t))] .$$

Corollary 7.3. $\mathcal{R}[C(t,\sigma)]$, $\sigma \leq t$ is monotone nondecreasing with decreasing σ .

Identical results hold with $C(t,\sigma)$ replaced by $\mathcal{D}(t,\sigma)$. Hence, if we denote the set of states controllable from (reachable at) time t by $\mathcal{P}c(t)(\mathcal{P}r(t))$ and denote the set of states which determinable at (observable from) time t by $\mathcal{Q}_d(t)(\mathcal{Q}_o(t))$, then there exist positive C^1 functions $\nu(t)$, $\omega(t)$, $\rho(t)$ such that

$$(7.6) \qquad \begin{cases} \mathcal{P}c(t) = \mathcal{R}[C(t,t+\mu(t))] \\ \mathcal{P}r(t) = \mathcal{R}[C(t,t-\nu(t))] \\ \mathcal{Q}_d(t) = \mathcal{R}[\mathcal{D}(t,t-\omega(t))] \\ \mathcal{Q}_o(t) = \mathcal{R}[\mathcal{D}(t,t+\rho(t))] \end{cases}$$

We can now characterize the concepts of controllability, reachability, determinability, and observability for the system (7.1).

Theorem 7.4. A state x_o of (7.1) is controllable from (reachable at) time τ if and only if there exists a finite value of time $t_1 > \tau$ ($t_1 < \tau$) such that $x_o \in R[C(\tau, t_1)]$.

Proof (for controllability only): (Sufficiency): The solution of (7.1) with initial state x_o and initial time τ is given by

$$x(t;\tau,x_o) = \Phi(t,\tau)x_o + \int_\tau^t \Phi(t,\eta)C(\eta)u(\eta)d\eta \ .$$

By hypothesis, there exists $z \in R^n$ such that $C(\tau,t_1)z = x_o$. Setting $u(\eta) = - C'(\eta)\Phi'(\tau,\eta)z$ and making use of the fact that $\Phi(t,\tau)$ satisfies a group property ($\Phi(t,\tau) = \Phi(t,\sigma)\Phi(\sigma,\tau)$ for all t, σ, τ) we find that $x(t_1;\tau,x_o) = 0$.

(Necessity): Suppose $x_o (\neq 0)$ is controllable from τ but $x_o \notin R[C(\tau,t)]$ for all $t > \tau$. Then there exists a finite value of time t_1 and an admissible control segment $u^*_{[\tau,t_1]}$ such that

$$0 = x(t_1;\tau,x_o) = \Phi(t_1,\tau)x_o + \int_\tau^{t_1} \Phi(t_1,\eta)C(\eta)u^*(\eta)d\eta$$

or

$$(7.7) \qquad x_o = - \int_\tau^{t_1} \Phi(\tau,\eta)C(\eta)u^*(\eta)d\eta \ .$$

Since $x_o \notin R[C(\tau,t_1)]$ it follows that x_o has a nonzero projection x_1 on $K[C(\tau,t_1)]$. By Theorem (7.1), $x_1 \in K[C(\tau,t)]$ for all $t \in [\tau,t_1]$ in which case $x_1'\Phi(\tau,\eta)C(\eta) = 0$ for all $\eta \in [\tau,t_1]$. But then, from (7.7), $x_1'x_o = 0$ which implies $x_o \in R[C(\tau,t_1)]$, a contradiction.

From Theorem 7.1 and Theorem 7.4 we deduce

Theorem 7.5. A system (7.1) is controllable from (reachable at) time τ if and only if there exists a finite value of time $t_1 > \tau$ ($t_1 < \tau$) such that rank $C(\tau,t_1) = n$.

Another useful result is given by

Theorem 7.6. (i) If $x \in Pc(t)$ and $\tau \leq t$, then $\Phi(\tau,t)\, x \in Pc(\tau)$.

(ii) If $x \in Pr(t)$ and $\tau \geq t$, then $\Phi(\tau,t)\, x \in Pr(\tau)$.

Proof of (i): By hypothesis there exists $t_1 > t$ and $u^*_{[t,t_1]}$ such that $x(t_1;t,x) = 0$. Then

$$0 = \Phi(t_1,t)x + \int_t^{t_1} \Phi(t_1,\eta)C(\eta)u^*(\eta)d\eta \tag{7.8}$$

$$= \Phi(t_1,\tau)\Phi(\tau,t)x + \Phi(t_1,\tau)\int_t^{t_1} \Phi(\tau,\eta)C(\eta)u^*(\eta)d\eta \quad .$$

Let

$$\hat{u}^* = u^* \quad \text{on} \quad [t,t_1]$$
$$= 0 \quad \text{on} \quad [\tau,t) \quad .$$

Then

$$x(t_1;\tau,\Phi(\tau,t)x) = \Phi(t_1,\tau)\Phi(\tau,t)x + \Phi(t_1,\tau)\int_\tau^t \Phi(\tau,\eta)C(\eta)\hat{u}^*(\eta)d\eta$$
$$= 0 \quad \text{by (7.8)}.$$

The proof of (ii) follows exactly as above. Now, from Definitions 7.3, 7.4 and Theorem 7.4, we have

Theorem 7.7. A state x_o of (7.1) is determinable at (observable from) time τ if and only if there exists a finite $t_1 < \tau$ $(t_1 > \tau)$ such that $x_o \in R[\mathcal{D}(\tau,t_1)]$.

Corollary 7.8. A system (7.1) is determinable at (observable from) time τ if and only if there exists a finite $t_1 < \tau$ $(t_1 > \tau)$ such that rank $\mathcal{D}(\tau,t_1) = n$.

The dual of Theorem 7.6 is also clear and is given by

Theorem 7.9. (i) If $x \in \mathcal{Q}_d(t)$ and $\tau \geq t$, then $\Phi'(t,\tau)\ x \in \mathcal{Q}_d(\tau)$.

(ii) If $x \in \mathcal{Q}_o(t)$ and $\tau \leq t$, then $\Phi'(t,\tau)\ x \in \mathcal{Q}_o(\tau)$.

The significance of the phrases "observable from" and "determinable at" should now be clear from the nature of (7.1). For if (7.1) is determinable at (observable from) time t, then (assuming $u(\cdot) \equiv 0$), the state of the system at time t can be uniquely determined from knowledge of the output $y(t)$ over a finite time interval ending at (beginning from) time t.

To see this, consider the solution for y of (7.1) when $u(\cdot) \equiv 0$, the initial state is x_o, the initial time is t_o; and assume (7.1) is determinable at t_o. Then we can write

$$y(t) = H(t)\Phi(t,t_o)x_o$$

and, for some $t_1 < t_o$,

$$x_o = \mathcal{D}(t_o,t_1)^{-1} \int_{t_o}^{t_1} \Phi'(t,t_o)H'(t)y(t)dt \; .$$

Application of Theorem 7.9 indicates that the state $\Phi'(t_o,t)x_o$ can be determined at time t.

It is also easy to check the following facts about the system (7.1).

1. Controllability from t implies that any state at time t can be transferred to any other state in a finite interval of time beginning with t.

2. If (7.1) is controllable from t and we reverse the ordering of the time scale, the reversed time system is reachable at t .

3. In general, complete controllability does not imply complete reachability or vice versa. To see this, consider (7.1) in which $n = p = r = 1$, $A(t) \equiv 0$, and

$$C(t) = \begin{cases} 1 - \cos t \,, & 0 \leq t \leq \frac{\pi}{t} \\ 0 \,, & t \leq 0 \\ 1 \,, & t \geq \frac{\pi}{2} \,. \end{cases}$$

The above system is completely controllable, but is reachable at t only for $t > 0$. One can say the following, however.

Proposition 7.10. (i) If a system (7.1) is controllable from time τ , then it is reachable at all $t \geq \tau + \mu(\tau)$ where $\mu(\cdot)$ is as defined in Corollary 7.2.

(ii) If a system is determinable at τ , it is observable from all $t \leq \tau - \omega(\tau)$ where $\omega(\tau)$ is as defined earlier.

Proof: (i) If (7.1) is controllable from τ , then

$$\text{rank } C(\tau,\tau + u(\tau)) = n = \text{rank } C(\tau + \mu(\tau),\tau)$$
$$= \text{rank } C(\xi,\tau) \quad \text{for all } \xi \geq \tau + \mu(\tau) \,.$$

(ii) Follows from (i) by duality

Corollary 7.11. Complete differential controllability (observability) is equivalent to complete differertial reachability (determinability). [It is therefore obvious that no distinction need be made between the concepts of controllability (observability) and reachability (determinability) for time-invariant systems (7.1)].

We now consider briefly the problem of controllability and determinability for ordinary nonlinear differential systems. The controllability problem can, in fact, be handled in a manner completely analogous to that presented for systems containing time delays (simply let $h \to 0$). Results of a slightly different nature can be obtained for special types of nonlinear equations and these are discussed in the section dealing with the application of Pfaffian systems to the controllability problem.

The determinability problem warrants more detailed comment, however, and so we consider the problem of giving sufficient conditions for a class of nonlinear ordinary differential systems to be uniformly determinable (defined below). The discussion is based on the work of Al'breckht and Krasovskii [1].

Consider the system

$$(7.9) \qquad \begin{cases} \dfrac{dx}{dt} = f(t,x(t)) \\ y(t) = g(t,x(t)) \end{cases}$$

where $x(t) \in R^n$, $y(t) \in R^r$, $f(t,0) \equiv 0$, f and g are analytic functions of x in a neighborhood of $x = 0$, and the components f_i and g_i of f and g respectively can be expanded in a series as follows.

$$(7.10)\qquad \begin{cases} f_i(t,x) = \sum_{m=1}^{\infty} \phi_i^{(m)}(t,x) \\ g_i(t,x) = \sum_{m=1}^{\infty} \psi_i^{(m)}(t,x) \end{cases}$$

where $\phi_i^{(m)}$ and $\psi_i^{(m)}$ are m^{th} order forms in x with continuous and bounded coefficients.

<u>Definition 7.6.</u> The system (7.9) is uniformly determinable at all $t \geq \gamma$ if there exists a function $Y : R^1 \times R^r \to R^n$ such that any trajectory $x(t)$ can be expressed as

$$(7.11)\qquad x(t) = Y(t,y(t+\theta)) \ , \ -\gamma \leq \theta \leq 0$$

for all $t \geq \gamma > 0$.

<u>Remark:</u> The interest in <u>uniform</u> determinability stems from the desire to develop a method of state-vector determination whose practical implementation would involve the taking of measurements over time intervals of <u>fixed</u> length.

The first result of interest characterizes uniform observability for the linear system (7.1) (with $u(\cdot) \equiv 0$) , and follows from Corollary 7.8 (See Kalman [17]).

<u>Lemma 7.10.</u> The system (7.1) is uniformly determinable at all $t \geq \gamma$ if and only if rank $\mathcal{D}(t,t-\gamma) = n$ for all $t \geq \gamma$, where

(7.12) $\mathcal{D}(t,t-\gamma) = \int_t^{t-\gamma} \Phi'(\eta,t)H'(\eta)H(\eta)\Phi(\eta,t)\,d\eta$

and Φ is the transition matrix corresponding to (7.3).

The main result is the theorem below.

Theorem 7.11. The system (7.9) is uniformly determinable at all $t \geq \gamma$ if its first variation has that property.

Proof: Consider the solution $x(\cdot,0,x_0)$ of (7.9) and suppose $x_0 \in N(0)$, a sufficiently small neighborhood of the origin in R^n. Writing $x(t) = x(t;0,x_0)$, we have the following expansion on the interval $[t-\gamma,t]$.

(7.3) $x(t+\theta) = \Phi(t+\theta,t)x(t) + \sum \beta^{(m)}(t,\theta,x(t)) , \quad -\gamma \leq \theta \leq 0$

where Φ is the transition matrix associated with the first variation of (7.9). The series (7.13) will converge for $x(t)$ sufficiently small. Since

(7.14) $y(t+\theta) = g(t+\theta\ , x(t+\theta)) , \quad -\gamma \leq \theta \leq 0$

then substituting (7.13) into (7.14) and expanding yields

$$y(t+\theta) = H(t+\theta)\Phi(t+\theta,t)x(t) + \sum_{m=2}^{\infty} \rho^{(m)}(t,\theta,x(t)) , \quad -\gamma \leq \theta \leq 0$$

where

$$H(t) = \frac{\partial g}{\partial x}(t,0) .$$

Now let

$$(7.15) \quad A(t) = Y(t,y(t+\theta)) = \mathcal{D}(t,t-\gamma)^{-1} \int_0^{-\gamma} \Phi'(t+\theta,t)H'(t+\theta)y(t+\theta)d\theta$$

$$= x(t) + \mathcal{D}(t,t-\gamma)^{-1} \int_0^{-\gamma} \Phi'(t+\theta,t)H'(t+\theta)\Big[\sum_{m=2}^{\infty} \rho^{(m)}(t,\theta,x(t))\Big]d\theta .$$

The above equation has the form $U(a;x,t) = 0$. The derivative of U with respect to x is 1 when a and x are set to zero. Hence, by the implicit function theorem, for any $a(t)$ in a sufficiently small neighborhood of the origin in R^n, there is a unique solution $x(t) = \pi(a(t),t)$ of (7.15). The map π is analytic in a in this nieghborhood so that $x(t)$ has a representation

$$x(t) = \mathcal{D}(t,t-\gamma)^{-1} \int_0^{-\gamma} \Phi'(t+\theta,t)H'(t+\theta)y(t+\theta)d\theta + \sum_{m=2}^{\infty} \alpha^{(m)}(t,a(t))$$

where $\alpha^{(m)}$ is an mth order form in a.

8. ALGEBRAIC CRITERIA FOR R^n CONTROLLABILITY OF LINEAR DELAY-DIFFERENTIAL AND ORDINARY-DIFFERENTIAL SYSTEMS

The conditions given thus far for controllability rest implicitly on the availability of the kernel matrix $K(t,\tau)$. From a computational point of view, especially in the case of delay equations, it is desirable to have controllability criteria which are solely dependent on the coefficients of the differential equation. Such criteria are developed in this section. The results extend some of those recently obtained by Kirillova and Čurakov [18].

The system model for the remaining discussion is (3.5) in which the coefficient matrices are assumed to possess $(n - 1)$ continuous derivatives.

The first theorem yields sufficient conditions for R^n - controllability. Define the matrix

$$(8.1)\quad Q(t) = [Q_1^1(t),\ldots,Q_1^n(t),Q_2^2(t - h),\ldots,Q_2^n(t - h),\ldots,Q_n^n(t - (n - 1)h)]$$

where

$$(8.2) \qquad Q_1^1(t) = C(t)$$

$$Q_j^{k+1}(t) = \frac{d}{dt} Q_j^k(t) - A(t + (j-1)h)Q_j^k(t) - B(t + (j-1)h)Q_{j-1}^k(t) ,$$

$$j = 1,\dots,k$$

$$k = 1,\dots,n$$

and $Q_j^k = 0$ for $j = 0$ or $j > k$.

Theorem 8.1. If there exists t_1 such that rank $Q(t_1) = n$, then (3.5) is R^n - controllable to the origin from some finite $t_o < t_1$.

Proof: We shall show that rank $Q(t_1) = n$ implies (5.1) . Fix t_1 and suppose (5.1) does not hold for any $t_o < t$. Then there exists a nonzero vector $z \in R^n$ such that $z'K(t_1,\tau)C(t) = 0$ for all $\tau \leqslant t_1$. In particular we have the set of equations

$$(8.3) \qquad z'K(t_1,\tau)C(\tau) = 0 \quad \text{for all} \quad \tau \in [t_1 - kh, t_1 - (k-1)h] ,$$

$$k = 0, 1,\dots,n \quad .$$

Differentiating $(n - 1)$ times we obtain, for $k = 0$,

$$(8.4)\quad z'[K(t_1,\tau)Q_1^i(\tau) + K(t_1,\tau + h)Q_2^i(\tau)] = 0\ ,\ \tau \in [t_1 - h,\ t_1]\ ,$$
$$i = 1,\ldots,n\ .$$

As $\tau \to t_1^-$, (8.4) becomes $z'Q_1^i(t_1) = 0\ ,\ i = 1,\ldots,n$

As $\tau \to (t_1 - h)^+$, (8.4) becomes $z'K(t_1,t_1 - h)Q_1^i(t_1 - h) = 0\ ,\ i = 1,\ldots,n\ .$

Setting $k = 1$ in (8.3) and differentiating $(n - 1)$ times again yields

$$(8.5)\quad z'[K(t_1,\tau)Q_1^i(\tau) + K(t_1,\tau + h)Q_2^i(\tau) + K(t_1,\tau + 2h)Q_3^i(\tau)] = 0\ ,$$
$$i = 1,\ldots,n\ .$$

As $\tau \to (t_1 - h)^-$, (8.5) becomes $z'Q_2^i(t_1 - h) = 0$

As $\tau \to (t_1 - 2h)^+$, (8.5) becomes $z'[K(t_1,t_1-2h)Q_1^i(t_1-2h)+K(t_1t_1-h)Q_2^i(t_1-h)] = 0.$

Continuing in this manner, for $k = n$ we obtain

$$z'Q_n^n(t_1 - (n - 1)h) = 0\ .$$

That is

$$z'Q_j^k(t_1 - (j - 1)h) = 0\ ,\ j = 1,\ldots,k\ ,\ k = 1,\ldots,n\ .$$

From this it follows that rank $Q(t_1) \neq n$ which proves the theorem.

Remarks: 1. For time-invariant systems with $\Lambda = 0$ the matrix Q becomes (8.6) $Q = [C, BC, \ldots, B^{n-1}C]$.

2. For time-varying systems with $B(\cdot) \equiv 0$, the matrix $Q(t)$ becomes

$$(8.6) \qquad Q(t) = [P_o(t), P_1(t), \ldots, P_{n-1}(t)] \quad \text{where}$$

$$P_o(t) = C(t) \;, \qquad P_k(t) = \frac{d}{dt} P_{k-1}(t) - A(t)P_{k-1}(t) \;.$$

To obtain necessary conditions for controllability of (3.5) is more difficult, and only partial results are available. We first present some results of Kirillova and Čurakov [18] for the case where A, B, C are constant matrices. Define the matrices P_j^r, $j = 1, \ldots, 2^{r-1}$, $r = 1, \ldots, n$, by

$$P_1^1 = C \;, \; P_{2j-1}^{r+1} = AP_j^r \;, \; P_{2j}^{r+1} = BP_j^r \;.$$

Let

$$(8.7) \qquad P = [P_1^1 \;, \; (P_{2j-1}^{r+1}) \;, \; (P_{2j}^{r+1})]_{j=1,\ldots,2^{r-1}}^{r=1,\ldots,n-1}$$

and let

$$P_m = [P_1^1 \;, \; (P_{2j-1}^{r+1}) \;, \; (P_{2j}^{r+1})]_{j=1,\ldots,2^{r-1}}^{r=1,\ldots,m-1}$$

Lemma 8.2: If rank $P = m < n$, then rank $P_m = m$.

Proof: Assume that for $j = 1,\ldots,s$, rank $P_j >$ rank P_k for all $k < j$ but rank $P_s =$ rank P_{s+1}. Then there exists a set of columns $v_1,\ldots,v_\sigma$, $s \leq \sigma \leq m$, of P_s which span the range of P_s. Let $V = [v_1,\ldots,v_\sigma]$. Then any matrix P_j^{s+1}, $j = 1,\ldots,2^s$ can be expressed as VK_1. Since $P_{2j-1}^{s+2} = AP_j^{s+1}$, $j = 1,\ldots,2^s$, we have that any matrix P_{2j-1}^{s+2} can be expressed as AVK_2. But it follows by hypothesis that AV has a representation VK_3. Then $\{v_1,\ldots,v_\sigma\}$ spans the range of P_{2j-1}^{s+2}. Similarly, since $P_{2j}^{s+1} = BP_j^s$, we can easily show that $\{v_1,\ldots,v_\sigma\}$ spans the range of P_{2j}^{s+2}. It then follows, by induction, that every column of P_{2j}^{k+1} and P_{2j-1}^{k+1}, $s < k \leq n-1$, is dependent on the v_k's. Therefore, $\sigma = m$, which proves the Lemma.

Theorem 8.3: A necessary condition for the time-invariant system (3.5) to be completely (differentially) R^n - controllable is that rank $P = n$.

Proof: Suppose rank $P = m < n$. Then there exists an $n \times n$ nonsingular matrix T such that

$$TP = \begin{bmatrix} \tilde{P} \\ 0 \end{bmatrix}$$

where $\tilde{P}$ has m rows and rank $\tilde{P} = m$. It follows from Lemma 8.2 that the coordinate transformation $x = Tx = \begin{bmatrix} x_1 \\ x_2 \end{bmatrix}$ transforms (3.5) into the form

$$(8.8)\qquad \begin{aligned} \dot{x}_1(t) &= A_{11}x_1(t) + A_{12}x_2(t) + B_{11}x_1(t-h) + B_{12}x_2(t-h) + C_1u(t) \\ \dot{x}_2(t) &= \qquad\qquad\quad A_{22}x_2(t) \qquad\qquad\qquad\qquad + B_{22}x_2(t-h) \;. \end{aligned}$$

It is obvious that (8.8) is not completely controllable.

It is pointed out in [18] that when A,B,C are constant in (3.5) and $n \leq 3$ then rank Q = rank P where Q is given by (8.1). Hence, for $n \leq 3$, Theorem 8.3 yields necessary <u>and</u> sufficient conditions for complete R^n - controllability of time-invariant systems (3.5). When $B = 0$, the matrix P becomes the well known controllability matrix, $[C,AC,\ldots,A^{n-1}C]$, for linear time-invariant systems.

It is somewhat unfortunate that what appears to be the natural extension of Lemma 8.2 to the time-varying case does not result in an immediately obvious corresponding extension of Theorem 8.3 except if the delay is zero, in which case, one can apply the following result due to Dolězal [9]. (See Weiss and Falb [25] for an alternate proof as well as further applications of the result to system theory).

<u>Lemma 8.4.</u> Let $S(t)$ be an $n \times n$ matrix of C^k functions defined on the real line. Let r be a nonnegative integer $< n$ such that rank $S(t) = r$ for all t. Then there exists an $n \times n$ matrix of C^k functions, $T(t)$, nonsingular for all t, such that

$$T(t)S(t) = \begin{bmatrix} \tilde{S}(t) \\ 0 \end{bmatrix}$$

where $\tilde{S}(t)$ is $r \times n$ and rank $\tilde{S}(t) = r$ for all t.

Now define the matrices

$$P_1^1(t) = C(t)$$

(8.9) $$P_{2j-1}^{r+1}(t) = \frac{d}{dt} P_j^r(t) - A(t + \alpha(j)h)P_j^r(t)$$

$$P_{2j}^{r+1}(t) = -B(t + \beta(j)h)P_j^r(t) \ .$$

where $r = 1,\dots,n-1$, and $j = 1,\dots,2^{r-1}$.

The functions $\alpha(\cdot)$ and $\beta(\cdot)$ are integral valued (both domain and range consist of natural numbers) and in an appendix to this section, we give an algorithm which yields the inverse image of each point in the range. Now let

$$P(t) = \left[P_1^1(t), (P_{2j-1}^{r+1}(t - \alpha(j)h)), (P_{2j}^{r+1}(t - \beta(j)h)) \right]_{j=1,\dots,2^{r-1}}^{r=1,\dots,n-1}$$

and let

$$\hat{P}_m(t) = \left[P_1^1(t), (P_{2j-1}^{r+1}(t - \alpha(j)h)), (P_{2j}^{r+1}(t - \beta(j)h)) \right]_{j=1,\dots,2^{r-1}}^{r=1,\dots,m-1} \ .$$

Lemma 8.5. Let I be any interval on which rank $\hat{P}_k(t)$ is a constant function of t for each k. If rank $P(t) = m$ for all $t \in I$ then rank $\hat{P}_m(t) = m$ for all $t \in I$.

(Remark: That such an interval I exists follows from the fact that a continuous matrix has unchanging rank on a union of intervals which form an everywhere dense set on the real line).

Proof: Assume that for $j = 1,\dots,s$, rank $\hat{P}_j(t) >$ rank $\hat{P}_r(t)$ for all $r < j$, all $t \in I$, but rank $\hat{P}_s(t) =$ rank $\hat{P}_{s+1}(t)$ for all $t \in I$.

Then, there exists on interval $J \subset I$ and a set of columns $w_1(t),\dots,w_\sigma(t)$, $s \le \sigma \le m$, of $P_s(t)$, which span the range of $\hat{P}_s(t)$ for each $t \in J$. Let $W(t) = [w_1(t),\dots,w_\sigma(t)]$. Then any maximal-row submatrix of any matrix $P_j^{s+1}(t - (\cdot)h)$, $j = 1,\dots,2^s$ has the form $W(t)K_1(t)$ for all $t \in J$. Since $P_{2j-1}^{s+2}(t) = \frac{d}{dt} P_j^{s+1}(t) - A(t + \alpha(j)h)P_j^{s+1}(t)$, $j = 1,\dots,2^s$, we have that any matrix $P_{2j-1}^{s+2}(t - \alpha(j)h)$ can be expressed as

$$\frac{d}{dt}(W(t)K_2(t)) - A(t)W(t)K_2(t)$$

$$= [\dot{W}(t) - A(t)W(t)]K_2(t) + \dot{W}(t)K_2(t) .$$

But it follows by hypothesis that $\dot{W}(t) - A(t)W(t)$ has a representation $W(t)K_3(t)$. Then $\{w_1(t),\dots,w_\sigma(t)\}$ spans the range of $P_{2j-1}^{s+2}(t - \alpha(j)h)$ at each $t \in J$. Similarly, since $P_{2j}^{s+1}(t) = B(t + \beta(j)h)P_j^s(t)$, it can be shown that $\{w_1(t),\dots,w_\sigma(t)\}$ spans the range of $P_{2j}^{s+2}(t - \beta(j)h)$ at each $t \in I]$. The Lemma then follows by induction, since rank $P(t) = m$.

Unfortunately, the presence of the delay term in (3.5) mitigates against an immediate generalization of Theorem (8.3) by means of a construction of the the type represented by (8.8). However, with the delay term absent $(h = 0)$, and with the simplification of setting $B(t) \equiv 0$, we obtain a result which, in conjunction with Lemma 8.4, yields a straightforward generalization of Theorem 8.3 and its proof for the case of ordinary differential equations. That is, let $h = 0$ and $B(t) \equiv 0$ in (3.5). With these assumptions, the matrices Q and P , in (8.1) and (8.9) respectively, coincide. And so, we have

Theorem 8.6: A necessary condition for the system (3.5) to be differentially controllable on an interval $[t_1,t_2]$ is that rank $P(t) = n$ for all $t \in [t_1,t_2]$ except possibly for a set whose complement is everywhere dense on $[t_1,t_2]$.

The direct proof of this result (See Silverman and Meadows [41]) is left as an exercise in light of Lemma 8.4 and the proof of Lemma 8.2. An indirect proof is obtained as a by-product of the discussion in the next chapter.

Corollary 8.7: A necessary and sufficient condition for a time-invariant system (7.1) to be completely controllable is that

$$\text{rank } [C, AC, \ldots, A^{n-1}C] = n \ . \tag{8.10}$$

Now consider (7.1) again and define the matrix $S(t) = [S_o(t), S_1(t), \ldots, S_{n-1}(t)]$ where $S_o(t) = H'(t)$, $S_k(t) = \frac{d}{dt} S_{k-1}(t) + A'(t)S_{k-1}(t)$. Then by duality we have

Corollary 8.8. A necessary and sufficient condition for a system (7.1) to be differentially observable (or determinable) on an interval $[t_1,t_2]$ is that rank $S(t) = n$ for all $[t_1,t_2]$ except possibly for a set whose complement is everywhere dense on $[t_1,t_2]$. [In the time invariant case, this reduces to rank $[H', A'H', \ldots, A^{n-1'}H'] = n$].

An application of Corollary 8.8 which follows directly from a proof of a theorem in [22] is that <u>for single-input, single-output systems, differential observability implies existence of an input/output differential equation defined on those intervals where rank</u> $S(t) = n$.

An interesting application of Corollary 8.7 is the following.

<u>Corollary 8.9.</u> The scalar time-invariant system

$$x^{(n)}(t) = u(t) \tag{8.11}$$

is completely differentially controllable.

<u>Proof:</u> The scalar system (8.11) is equivalent to the time-invariant n - dimensional system (7.1) where

$$A = \begin{bmatrix} 0 & I_{n-1} \\ & \\ 0 & 0 \end{bmatrix} \qquad C = \begin{bmatrix} 0 \\ \vdots \\ 0 \\ 1 \end{bmatrix}$$

and I_{n-1} is the $(n-1) \times (n-1)$ identity. Then (8.10) holds. Corollary 8.9 can be generalized as follows. Consider (8.11) as a delay equation with initial data available for $t \leq t_o$. Define $L_t(\cdot)$ as any (nonlinear, time-varying) functional operator such that trajectories of (8.11) are in the domain of $L_t(\cdot)$, and consider the system

$$x^{(n)}(t) + L_t(x)(t) = u(t) . \tag{8.12}$$

As an example we might consider

$$L_t(x) = f(t, x(t), x^{(1)}(t) \ldots, x^{(n)}(t), x(t-h_1) x^{(1)}(t-h_2) \ldots, x^{(n)}(t-h_n)).$$

<u>Theorem 8.10</u> The system (8.12) is completely differentially R^n - controllable.

<u>Proof:</u> Consider (8.11) with arbitrary initial condition x_o and let the control function which transfers x_o to the origin in some arbitrarily given time be u^* with the associated trajectory denoted by x^*. Then the system (8.12) with initial function ϕ such that $\phi(t_o) = x_o$ can also be steered to the origin in R^n along the trajectory x^* by applying the control $u^*_{(t)} + L_t(x^*)(t)$.

This result which generalizes a result of Hermes [13] and of Davison et al [7], can be simply interpreted as <u>differential controllability of scalar n^{th} order linear differential systems is invariant under feedback</u>. We shall return to (8.12) for further discussion in the next section.

L. Weiss

APPENDIX

ALGORITHM FOR GENERATING THE FUNCTIONS $\alpha(j)$ AND $\beta(j)$.

To generate $\alpha(j)$:

Let ℓ be a nonnegative integer and let X_e be the set of (odd) integers $\{x_{ek}\}$ (ordered by >) such that $\alpha(j) = \ell$ corresponds to $2j - 1 = x_{\ell k}$, $k = 1,\ldots,\frac{(n-2)!}{(n-(\ell+2))!\ell!}$.

Then

$$X_\ell = \bigcup_{\mu=1}^{n-(\ell+1)} \left\{2^\mu x_{\ell-1,k} + 1 \ , \ k = 1,\ldots,\frac{(n-(\mu+2))!}{(n-(\ell+\mu+1))!\ (\ell-1)!}\right\}$$

where $x_{-1,k} \equiv 0$.

To generate $\beta(j)$:

Let ℓ be a positive integer and let Y_ℓ be the set of (even) integers $\{y_{\ell k}\}$ (ordered by >) such that $\beta(j) = \ell$ corresponds to $2j = y_{\ell k}$, $k = 1,\ldots,\frac{(n-2)!}{(n-(\ell+1))!\ (\ell-1)!}$.

Then

$$Y_\ell = \bigcup_{\mu=1}^{n-\ell} \left\{2^\mu y_{\ell-1,k} - (2^\mu - 2) \ , \ k = 1,\ldots,\frac{(n-(\mu+2))!}{(n-(\ell+\mu))!\ (\ell-2)!}\right\}$$

where $y_{o,k} \equiv 1$.

9. THE PFAFFIAN SYSTEM APPROACH TO CONTROLLABILITY

In this section, we present an alternate approach to the problem of characterizing differential controllability of linear ordinary differential systems, and local controllability of nonlinear systems with control appearing linearly. Our discussion stems from the work of Hermes [13], which was, in turn, motivated by the work of Caratheodory [3] and Chow [4]. A novel aspect is the connection of Pfaffians to Theorem 8.6.

Consider the system (7.1) in which $p < n$ and $u(\cdot) \varepsilon L_2$. Suppose rank $C(t) = s =$ constant on an open interval I.

Let $R(t)$ be an $(n - s) \times n$ matrix of continuous functions defined on I, with rank $R(t) = n - s$ for all $t \varepsilon I$, such that

(9.1) $$R(t)\ C(t) = 0 \text{ for all } t \varepsilon I .$$

Definition 9.1. The Pfaffian system associated with (9.1) is the system

(9.2) $$R(t)\ dx - R(t)\ A(t)\ x\ dt = 0 \text{, defined on } I .$$

Let the rows of R be denoted by r_i' and let r' denote an arbitrary nonzero linear combination of those rows with continuous scalar valued coefficients $\alpha_i(t)$. That is

(9.3) $$r'(t) = \sum_i \alpha_i(t) r_i'(t) .$$

<u>Definition 9.2.</u> The Pfaffian system (9.2) is integrable at the point $t_1 \varepsilon I$ if there exists some $r'(t)$ such that the form

(9.4) $$r'(t)\, dx - r'(t)\, A(t)\, x\, dt$$

is an exact differential in a neighborhood of t_1.

More precisely, integrability of the Pfaffian at t_1 implies existence of a C^1 scalar-valued function $\psi(t,x)$ such that for some $\varepsilon > 0$,

$$\frac{\partial\psi}{\partial x}(t,x) = r'(t) \quad ; \quad \frac{\partial\psi}{\partial t}(t,x) = -r'(t)\, A(t)\, x$$

for $t \varepsilon [t_1, t_1 + \varepsilon)$, where $\frac{\partial\psi}{\partial x} = \text{row}\, (\frac{\partial\psi}{\partial x_i})$.

<u>Remark</u>: Hermes [13] shows that any integrating factor of (9.4) can be taken as a function only of t and can therefore be incorporated in (9.3).

The main result we wish to prove is the following.

<u>Theorem 9.1</u>. Consider (7.1) and its associated Pfaffian (9.3). Let the matrices A, C possess $(n - 2)$ and $(n - 1)$ continuous derivatives respectively and define the matrix $Q(t)$ as in (8.6). Then the Pfaffian is nonintegrable on an interval $[t_1,t_2]$ if and only if rank $Q(t) = n$ for all t on a subset of $[t_1,t_2]$ which is everywhere dense on $[t_1,t_2]$.

<u>Proof:</u> (Sufficiency): Suppose there exists $\tau \in [t_1, t_2]$ such that (9.2) is defined and integrable at τ (so rank $C(t)$ = constant on an open interval which includes τ). Then there exists r' , a nonzero linear combination of the rows of R , and $\varepsilon > 0$ with $\tau + \varepsilon \leqslant t_2$ such that $\dot{r}'(t) = -r'(t)A(t)$ for $t \in [\tau, \tau + \varepsilon)$. Since $r'(t)C(t) = 0$ on the latter interval, we have $\dot{r}'(t)C(t) + r'(t)\dot{C}(t) = 0$ on the interval which implies $r'(t)Q(t) = 0$ on $[\tau, \tau + \varepsilon)$. It then follows that a subinterval $J \subset [\tau, \tau + \varepsilon)$ exists such that rank $Q(t) < n$ on J . The contrapositive of this establishes sufficiency.

(Necessity): Suppose there exists $J \subset [t_1, t_2]$ such that rank $Q(t) < n$ on J . Then there exists $I \subset J$ such that rank $Q(t) = m < n$ for all $t \in I$, and rank $C(t) = s \leqslant m$ for all $t \in I$, where m, s are constants. By Lemma 8.4 there exists an $n \times n$ matrix of C^1 functions, $T(t)$, defined and nonsingular for all $t \in I$, such that

$$T(t)Q(t) = \begin{bmatrix} Q_1(t) \\ 0 \end{bmatrix} \quad \text{and} \quad T(t)C(t) = \begin{bmatrix} C_1(t) \\ 0 \end{bmatrix}$$

where $Q_1(t)$ and $C_1(t)$ have m rows and rank $Q_1(t) = m$ for all $t \in I$. Now since rank $C_1(t) = s$ for all $t \in I$, then by Lemma 8.4, there exists an $m \times m$ nonsingular C^1 matrix $T_1(t)$ such that

$$\begin{bmatrix} T_1(t) & 0 \\ 0 & I_{n-} \end{bmatrix} \begin{bmatrix} C_1(t) \\ 0 \end{bmatrix} \triangleq T_2(t) \begin{bmatrix} C_1(t) \\ 0 \end{bmatrix} = \begin{bmatrix} C_2(t) \\ 0 \end{bmatrix}$$

where $C_2(t)$ has s rows and rank $C_2(t) = s$ on I . Let $K = [0 \quad I_{n-s}]$ be an $(n-s) \times n$ matrix. Then the matrix $R(t) = KT_2(t)T(t)$ is $(n-s) \times n$, has rank $n - s$ on I , and $R(t)C(t) \equiv 0$ on I . Let

$\beta' = [0,\ldots,0,1]$ (n-dimensional) and for fixed $\eta\varepsilon I$, let $r'(t) = \beta' T(\eta)\Phi(\eta,t)$ where Φ is the transition matrix associated with (7.1). Then $r'(t)Q(t) = 0$ on I . Moreover, $r'(t)C(t) = 0$ and $\dot{r}'(t) = -r'(t)A(t)$ for all $t\varepsilon I$. Define the scalar valued function $\psi(t,x) = r'(t)x$. Then, for $t\varepsilon I$,

$$\frac{\partial\psi}{\partial x}(t,x) = r'(t) \quad ; \quad \frac{\partial\psi}{\partial t}(t,x) = -r'(t)A(t)x$$

which implies existence of $\tau\varepsilon I$ such that the Pfaffian is integrable at τ . The contrapositive of what we have proved establishes necessity and proves the theorem.

Now, the following result was proved by Hermes [13] using the controllability matrix (7.4).

Theorem 9.2. A system (9.1) is differentially controllable on an interval $[t_1,t_2]$ if and only if the Pfaffian is nonintegrable on $[t_1,t_2]$.

The proof of Theorem 8.6 now follows trivially from Theorems 9.1 and 9.2.

We discuss briefly the applicability of this method to nonlinear systems with control appearing linearly. The modifications in this case are relatively minor for ordinary differential equations. That is, if the system is of the form

$$\dot{x} = f(t,x) + C(t,x)u \ , \ (t,x) \ \varepsilon \ \mathcal{D} \ \text{(a domain in } R^n) \ .$$

Then (9.2), (9.3), (9.4) becomes, respectively

$$R(t,x)C(t,x) = 0 \quad \text{for all} \quad (t,x) \in \mathcal{D}$$

$$R(t,x)dx - R(t,x)f(t,x)dt = 0 \quad \text{for all} \quad (t,x) \in \mathcal{D}$$

$$r'(t,x) = \sum_i \alpha_i(t,x)r_i'(t,x)$$

and Definition 9.2 is altered accordingly [we now speak of integrability at the point $(t_1,x_1)\in\mathcal{D}$ and of a neighborhood of (t_1x_1)], although its basic structure is unchanged. Now, if $f(t,x)$ is replaced by some functional operator $M_t(x)$ (which may introduce delays, etc.), structure is unchanged. Now, if $f(t,x)$ is replaced by some functional operator $M_t(x)$ (which may introduce delays, etc.), the situation becomes more complicated with respect to the definition of the Pfaffian and its integrability. However, under some circumstances, the definition given for ordinary differential equations is formally applicable because of the special nature of $M_t(x)$. We now give an example of this, by adapting an example given in [13].

Consider the scalar n^{th} - order system (8.10). The system is equivalent to the n - dimensional system

$$\text{(9.5)} \qquad \begin{bmatrix} \dot{x}_1 \\ \cdot \\ \cdot \\ \cdot \\ \cdot \\ \dot{x}_{n-1} \end{bmatrix} = \begin{bmatrix} 0 & I_{n-1} \end{bmatrix} \begin{bmatrix} x_1 \\ \cdot \\ \cdot \\ \cdot \\ \cdot \\ x_n \end{bmatrix}$$

$$\dot{x}_n = -L_t(x_1 \ldots x_n) + u$$

where I_{n-1} is the $(n-1)\times(n-1)$ identity. Except for the last equation, (9.5) is of a form for which Definitions 9.1 and 9.2 make sense. Clearly, if R can be chosen so that L_t is excluded from entering the expression for the Pfaffian, then the results in this section become applicable to (9.5) and therefore to (8.10). It turns out that R can be so chosen and we therefore have

Proposition 9.3. The Pfaffian of the system (9.6) is nonintegrable for all t.

Proof: Proceeding formally, we choose $R(t)$ as the $(n - 1) \times n$ matrix

$$R(t) = [I_{n-1} \; 0] .$$

The Pfaffian system associated with (9.5) is

$$dx_i - x_{i+1}\, dt = 0 \; , \; i = 1,\dots,n - 1 . \tag{9.6}$$

For (9.6) to be integrable, there must exist scalar-valued functions $\alpha_i(t)$, not all zero, such that

$$\begin{bmatrix} 0 & \alpha_{n-1} \cdots \alpha_1 \end{bmatrix} \begin{bmatrix} dx_n \\ \vdots \\ dx_1 \end{bmatrix} - \begin{bmatrix} \alpha_{n-1} \cdots \alpha_1 & 0 \end{bmatrix} \begin{bmatrix} x_n \\ \vdots \\ x_1 \end{bmatrix} dt$$

is an exact differential. But then we would have

$$[0\ \dot{\alpha}_{n-1} \dots \dot{\alpha}_1] = [\alpha_{n-1} \dots \alpha_1\ 0]$$

from which it follows that $\alpha_i(t) \equiv 0$, $i = 1,\dots,n-1$. Hence the Pfaffian is nonintegrable.

What we have essentially shown is that <u>nonintegrability of the Pfaffian for scalar n^{th} - order system is invariant under feedback</u>.

In his paper, Hermes argued that on the basis of results for linear systems it was reasonable to define controllability for nonlinear systems with control appearing linearly in terms of non-integrability of the Pfaffian for such systems. Since we proved earlier that the system (8.10) is completely differentially (R^n -) controllable, then we can state that every system which is presently known to be completely differentially (R^n -) controllable has a nonintegrable Pfaffian associated with it. At this point it appears, therefore, that Hermes assertion is correct provided the word "controllability" is modified to "differential controllability".

10. STRUCTURE THEORY FOR LINEAR DIFFERENTIAL SYSTEMS

For a number of years following the introduction of the concept of "state" as a fundamental quality in system theory and control theory, there existed a certain amount of confusion in the literature concerning the relationship among different mathematical models or representations of the same system. For instance, there was the transfer function representation, the impulse response , the input/output differential equation, and the state vector differential equation. In the case of linear time-invariant systems described by ordinary differential equations, this confusion was cleared up primarily by R. E. Kalman through the statement and proof of a fundamental theorem on the structure of linear time-invariant control systems [16] which was motivated by the significant earlier work of E. G. Gilbert [11].

We shall develop the structure theory within the more general context of time-varying systems (See Weiss [14]) and discuss some of its implications.

Our efforts in this section are limited to a discussion of linear ordinary differential systems of the form (7.1). For convenience, (7.1) is sometimes referred to as "the system $\{A(\cdot), C(\cdot), H(\cdot)\}$".

A principal tool in our development of this part of the structure theory is the following Corollary to Lemma 8.4.

<u>Corollary 10.1</u>: Let $S(t)$ be as in Lemma 8.4 with the additional property that it is symmetric. Then there exists an $n \times n$ matrix of C^k functions $T(t)$, nonsingular for all t, such that

$$(10.1)\quad T(t)S(t)T'(t) = \begin{bmatrix} \tilde{S}(t) & 0 \\ 0 & 0 \end{bmatrix} \quad \text{for all } t$$

where $\tilde{S}(t)$ is $r \times r$ and rank $\tilde{S}(t) = r$ for all t.

<u>Theorem 10.2.</u> Consider the system (7.1) with controllability matrix $C(t,t+\mu(t))$ and suppose rank $C(t,t+\mu(t)) = r_c < n$ for all t. Then there exists a C^1 invertible coordinate transformation of the state space of (7.1) with respect to which (7.1) takes on the form

$$\dot{x}_1(t) = A_{11}(t)x_1(t) + A_{12}(t)x_2(t) + C_1(t)u(t)$$

$$(10.2)\qquad \dot{x}_2(t) = A_{22}(t)x_2(t)$$

$$y(t) = H_1(t)x_1(t) + H_2(t)x_2(t)$$

valid for all time, where $x_1(t)$ is an r_c - vector. Moreover, the system $\{A_{11}(\cdot), C_1(\cdot), H_1(\cdot)\}$ is completely controllable.

<u>Proof:</u> Application of Corollary 10.1 to $C(t,t+\mu(t))$ shows existence of a continuously differentiable $n \times n$ matrix, $T(t)$, nonsingular for all t, such that

$$(10.3)\qquad T(t)C(t,t+\mu(t))T'(t) = \begin{bmatrix} \tilde{C}(t) & 0 \\ 0 & 0 \end{bmatrix}$$

where $\tilde{C}(t)$ is $r_c \times r_c$, symmetric, and rank $\tilde{C}(t) = r_c$ for all t. The right side of (10.3) represents the controllability matrix for (7.1) after the transformation $\tilde{x}(t) = T(t)x(t)$ is made. Hence, by Theorem 7.4, a controllable state in the transformed system has the form $\begin{matrix} \tilde{x}_1 \\ 0 \end{matrix}$ where $\tilde{x}_1$ is an r_c - vector. From Theorem 7.6 we have that the transformed transition matrix $\tilde{\Phi}$ has the form (independent of arguments)

$$(10.4)\qquad \tilde{\Phi} = \begin{bmatrix} \Phi_{11} & \Phi_{12} \\ 0 & \Phi_{22} \end{bmatrix}$$

where Φ_{11} is $r_c \times r_c$. It then follows by equation (7.3) that regardless of t, $\tilde{A}$ has the form

$$\tilde{A} = \begin{bmatrix} A_{11} & A_{12} \\ 0 & A_{22} \end{bmatrix} .$$

The above transformed quantities are related to the original by the equations

$$\tilde{\Phi}(t,\tau) = T(t)\Phi(t,\tau)T^{-1}(\tau)$$

$$\tilde{A}(t) = T(t)A(t)T^{-1}(t) + \dot{T}(t)T^{-1}(t) .$$

Also, (10.3) implies that

$$T(t)\Phi(t,\eta)C(\eta)C'(\eta)\Phi'(t,\eta)T'(t) = \begin{bmatrix} K(t,\eta)K'(t,\eta) & 0 \\ 0 & C \end{bmatrix}$$

for all t and all $\eta \varepsilon [t, t + \mu(t)]$ where $K(t,\eta)$ is $r_c \times p$. Choosing $\eta = t$, it is clear that the above equation implies

$$T(t)C(t) = \begin{bmatrix} K(t,t) \\ 0 \end{bmatrix} \quad \text{for all } t .$$

Using the notation

$$\tilde{C}(t) = T(t)C(t) = \begin{bmatrix} C_1(t) \\ C_2(t) \end{bmatrix}$$

where $C_1(t)$ is $r_c \times p$, we see that $K(t,t) = C_1(t)$ and $C_2(t) = 0$ for all t which proves the main part of the theorem. The remainder follows by a trivial observation.

By completely analogous argument, one can prove

Theorem 10.3: Consider (1) with determinability matrix $\mathcal{D}(t,t - \omega(t))$ and let rank $\mathcal{D}(t,t - \omega(t) = r_d < n$ for all t. Then there exists a C^1 invertible coordinate transformation of (7.1) such that under this transformation

$$
(10.5) \qquad \begin{cases} \dot{x}_1(t) = A_{11}(t)x_1(t) \ C_1(t)u(t) \\ \dot{x}_2(t) = A_{21}(t)x_1(t) + A_{22}(t)x_2(t) + C_2(t)u(t) \\ y(t) = H_1(t)x_1(t) \end{cases}
$$

valid for all t where $x_1(t)$ is an r_d - vector. Furthermore, the system $\{A_{11}(\cdot), C_1(\cdot), H_1(\cdot)\}$ is completely determinable.

Theorem 10.4. Consider (7.1) and let the hypotheses of Theorem 10.2 hold so that (7.1) can be transformed into (10.2) with the transformed transition matrix given by (10.4). Define the determinability matrices $\mathcal{D}_1$ and $\mathcal{D}_2$ by

$$
(10.6) \qquad \mathcal{D}_1(t,\sigma) = \int_t^\sigma \Phi'_{11}(\eta,t)H_1'(\eta)H_1(\eta)\Phi_{11}(\eta,t)d\eta
$$

$$
(10.7) \qquad \mathcal{D}_2(t, \ ) = \int_t^\sigma \Phi'_{22}(\eta,t)H_2'(\eta)H_2(\eta) \ {}_{22}(\eta,t)d\eta
$$

and let $\omega_i(t)$, $i = 1,2$ be C^1 functions that

$$
R[\mathcal{D}_i(t,t - \omega_i(t)] = \bigcup_{\sigma<t} R[\mathcal{D}_i(t,\sigma)] \ , \ i = 1,2 \ ,
$$

and let

$$\operatorname{rank}\ \mathcal{D}_1(t, t-\omega_1(t)) = r_{d_1} < r_c \quad \text{for all } t$$

$$\operatorname{rank}\ \mathcal{D}_2(t, t-\omega_2(t)) = r_{d_2} < n - r_c \quad \text{for all } t .$$

Then there exists a C^1 invertible coordinate transformation, $\hat{x}(t) = T(t)x(t)$, defined for all t, under which the coefficient matrices of (7.1) have the form

$$A(t) \rightarrow \begin{bmatrix} A^{aa}(t) & 0 & A^{ac}(t) & A^{ad}(t) \\ A^{ba}(t) & A^{bb}(t) & A^{bc}(t) & A^{bd}(t) \\ 0 & 0 & A^{cc}(t) & 0 \\ 0 & 0 & A^{dc}(t) & A^{dd}(t) \end{bmatrix}$$

(10.8)

$$C(t) \rightarrow \begin{bmatrix} C^a(t) \\ C^b(t) \\ 0 \\ 0 \end{bmatrix} \quad ; \quad H(t) \rightarrow [H^a(t) \quad 0 \quad H^c(t) \quad 0]$$

<u>Proof:</u> Let $T_1(t)$ be the diffeomorphism which transforms (7.1) into (10.2). With the resulting transition matrix given by (10.4), the determinability matrix for the transformed system is (with arguments omitted)

$$\mathcal{D}(t,\sigma) = \int_t^{\sigma} \begin{bmatrix} \Phi'_{11}H'_1H_1\Phi_{11} & \Phi'_{11}H'_1H_1\Phi_{12} + \Phi'_{11}H'_1H_2\Phi_{22} \\ \Phi'_{12}H'_1H_1\Phi_{11} + \Phi'_{22}H'_2H'_1\Phi_{11} & \Phi'_{12}H'_1H_1\Phi_{12} + \Phi'_{12}H'_1H_2\Phi_{22} + \\ & + \Phi'_{22}H'_2H_1\Phi_{12} + \Phi'_{22}H'_2H_2\Phi_{22} \end{bmatrix} ds$$

$$= \begin{bmatrix} \mathcal{D}_{11} & \mathcal{D}_{12} \\ \mathcal{D}_{21} & \mathcal{D}_{22} \end{bmatrix} .$$

From (10.6) and (10.7) we have $\mathcal{D}_{11} = \mathcal{D}_1$ and the last term in $\mathcal{D}_{22}$ is $\mathcal{D}_2$. Clearly, $\mathcal{D}_1$ is the determinability matrix for the system $\{A_{11}(\cdot), C_1(\cdot), H_1(\cdot)\}$ in (10.2). By hypothesis, rank $\mathcal{D}_1(t, t-\omega_1(t)) = r_{d_1} < r_c$ for all t. Hence, by Corollary 10.1 there exists a continuously differentiable $r_c \times r_c$ nonsingular matrix $T_2(t)$ such that

$$T'_2(t)\mathcal{D}_1(t, t-\omega_1(t))T_2(t) = \begin{bmatrix} \hat{\mathcal{D}}_1(t) & 0 \\ 0 & 0 \end{bmatrix} \quad \text{for all } t$$

where $\hat{\mathcal{D}}_1(t)$ is $r_{d_1} \times r_{d_1}$ and rank $\hat{\mathcal{D}}_1(t) = r_{d_1}$ for all t.

Hence, by Theorem 10.3 the transformation

$$\begin{bmatrix} x^a(t) \\ x^b(t) \end{bmatrix} = T_2(t)^{-1}x_1(t)$$

transforms (7.1) into a form in which

$$A_{11}(t) \to \begin{bmatrix} A^{aa}(t) & 0 \\ A^{ba}(t) & A^{bb}(t) \end{bmatrix}$$

$$C_1(t) \to \begin{bmatrix} C^a(t) \\ C^b(t) \end{bmatrix} \quad ; \; H_1(t) \to [H^a(t) \quad 0] \; .$$

Thus, (7.1) becomes

$$\begin{bmatrix} \dot{x}^a(t) \\ \dot{x}^b(t) \end{bmatrix} = \begin{bmatrix} A^{aa}(t) & 0 \\ A^{ba} & A^{bb}(t) \end{bmatrix} \begin{bmatrix} x^a(t) \\ x^b(t) \end{bmatrix} + T_2(t)^{-1}A_{12}(t)x_2(t)$$

$$+ \begin{bmatrix} C^a(t) \\ C^b(t) \end{bmatrix} u(t) \tag{10.9}$$

$$\dot{x}_2(t) = A_{22}(t)x_2(t)$$

$$y(t) = H^a(t)x^a(t) + H_2(t)x_2(t)$$

where we have $\dim x^a(t) = r_{d_1}$ and $\dim x^b(t) = r_c - r_{d_1}$.

Now consider the system $\{A_{22}(\cdot),\ 0\ H_2(\cdot)\}$ in (10.9). The determinability matrix for this system is $\mathcal{D}_2$. By hypothesis, rank $\mathcal{D}_2(t,t - \omega_2(t)) = r_{d_2} < n - r_c$ for all t . Then there exists an $(n - r_c) \times (n - r_c)$ continuously differentiable, nonsingular matrix $T_3(t)$ such

$$T_3'(t)\mathcal{D}_2(t,t-\omega_2(t))T_3(t) = \begin{bmatrix} \hat{\mathcal{D}}_2(t) & 0 \\ 0 & 0 \end{bmatrix} \quad \text{for all } t .$$

where $\hat{\mathcal{D}}_2(t)$ is $r_{d_2} \times r_{d_2}$ and rank $\mathcal{D}_2(t) = r_{d_2}$ for all t.

The coordinate transformation defined by

$$T_3(t)^{-1} x_2(t) = \begin{bmatrix} x^c(t) \\ x^d(t) \end{bmatrix}$$

transforms (10.9) into the desired canonical form in which

$$T_2(t)^{-1}A_{12}(t)T_3(t) = \begin{bmatrix} A^{ac}(t) & A^{ad}(t) \\ A^{bc}(t) & A^{bd}(t) \end{bmatrix}$$

$$\dot{T}_3(t)^{-1}T_3(t) + T_3(t)^{-1}A_{22}(t)T_3(t) = \begin{bmatrix} A^{cc}(t) & 0 \\ A^{dc}(t) & A^{dd}(t) \end{bmatrix}$$

$$H_2(t)T_3(t) = \begin{bmatrix} H^c(t) & 0 \end{bmatrix}$$

i.e., the system now becomes

$$(10.10) \begin{cases} \dot{x}^a(t) = A^{aa}(t)x^a(t) + A^{ac}(t)x^c(t) + A^{ad}(t)x^d(t) + C^a(t)u(t) \\ \dot{x}^b(t) = A^{ba}(t)x^a(t) + A^{bb}(t)x^b(t) + A^{bc}(t)x^c(t) + A^{bd}(t)x^d(t) + C^b(t)u(t) \\ \dot{x}^c(t) = A^{cc}(t)x^c(t) \\ \dot{x}^d(t) = A^{dc}(t)x^c(t) + A^{dd}(t)x^d(t) \\ y(t) = H^a(t)x^a(t) + H^c(t)x^c(t) \end{cases}$$

valid for all t where $\dim x^c(t) = r_{d_2}$ and $\dim x^d(t) = n - r_c - r_{d_2}$. This completes the proof of Theorem 10.3 in which the overall coordinate transformation $T(t)$ is given by

$$T(t) = \begin{bmatrix} T_2(t)^{-1} & 0 \\ 0 & T_3(t)^{-1} \end{bmatrix} T_1(t) .$$

Our final theorem in this section, when taken together with Theorems 10.2, 10.3, 10.4, yields the structural decomposition of a given system (7.1). It is motivated by the possibility that certain state variables in x^d may be determinable as a result of the connection A^{ad}, i.e., the system associated with $\begin{bmatrix} x^a \\ x^d \end{bmatrix}$ may contain a determinable subsystem.

Theorem 10.5: Consider the system (7.1) and let the hypotheses of Theorem 10.4 hold so that (7.1) can be transformed into (10.10). Consider the system

$$(10.11) \qquad \begin{bmatrix} A^{aa}(\cdot) & A^{ad}(\cdot) \\ 0 & A^{dd}(\cdot) \end{bmatrix}, \quad \begin{bmatrix} C^a(\cdot) \\ 0 \end{bmatrix}, \quad \begin{bmatrix} H^a(\cdot) & 0 \end{bmatrix}$$

and call the corresponding determinability matrix $\mathcal{D}_3(t,\sigma)$. Let $\omega_3(t)$ be a C^1 function such that

$R[\mathcal{D}_3(t,t-\omega_3(t)] = \bigcup_{\sigma<t} R[\mathcal{D}_3(t,\sigma)]$, and let rank $\mathcal{D}_3(t,t-\omega_3(t)) = r_{d_3}$ for all t , $r_{d_3} < n - r_c - r_{d_2} + r_{d_1}$. Then there exists a C^1 invertible coordinate transformation which converts (10.10) into the form

$$
\begin{aligned}
\dot{x}^a(t) &= A^{aa}(t)x^a(t) + A^{ac}(t)x^c(t) + C^a(t)u(t) \\
\dot{x}^b(t) &= A^{ba}(t)x^a(t) + A^{bb}(t)x^b(t) + A^{bc}(t)x^c(t) + A^{bd}(t)x^d(t) + \\
&\qquad + C^b(t)u(t) \qquad\qquad (10.12) \\
\dot{x}^c(t) &= A^{cc}(t)x^c(t) \\
\dot{x}^d(t) &= A^{dc}(t)x^c(t) + A^{dd}(t)x^d(t) \\
y(t) &= H^a(t)x^a(t) + H^c(t)x^c(t)
\end{aligned}
$$

valid for all time, where

$\dim x^a(t) = r_{d_1}$; $\dim x^b(t) = r_c - r_{d_1}$; $\dim x^c(t) = r_{d_2} + r_{d_3} - r_{d_1}$; $\dim x^d(t) = n - r_c - r_{d_2} + r_{d_1}$.

Note: The general form differs from (10.10) in that, by means of a further transformation of coordinates plus a regrouping of state variables ($\dim x^c(t) > \dim x^c(t)$ and $\dim x^d(t) < x^d(t)$), the feedback coefficient from the system associated with x^d to that associated with x^a becomes identically zero.

<u>Proof of Theorem 10.5:</u> Let the transition matrix for the system (10.11) be given by

$$\begin{bmatrix} \Phi^{aa} & \Phi^{ad} \\ 0 & \Phi^{dd} \end{bmatrix}$$

where Φ^{aa} corresponds to A^{aa} and has dimension $r_{d_1} \times r_{d_1}$. Then omitting arguments in the integrand below, we have

$$\mathcal{D}_3(t,\sigma) = \int_t^{\sigma} \begin{bmatrix} \Phi^{aa'} H^{a'} H^{a} \Phi^{aa} & \Phi^{aa'} H^{a'} H^{a} \Phi^{ad} \\ \Phi^{ad'} H^{a'} H^{a} \Phi^{aa} & \Phi^{ad'} H^{a'} H^{a} \Phi^{ad} \end{bmatrix} d\eta$$

or

$$(10.13) \qquad \mathcal{D}_3(t,\sigma) = \begin{bmatrix} \mathcal{D}^{aa}(t,\sigma) & R(t,\sigma) \\ R'(t,\sigma) & Q(t,\sigma) \end{bmatrix}$$

where $\mathcal{D}^{aa}(t,\sigma)$ is the determinability matrix for $\{A^{aa}(\cdot), C^{a}(\cdot), H^{a}(\cdot)\}$, so that rank $\mathcal{D}^{aa}(t,t-\omega_1(t)) = r_{d_1}$ for all t. Let $\omega_4(t)$ be a continuously differentiable function such that

$$R[Q(t,t-\omega_4(t)] = \bigcup_{\sigma<t} R[Q(t,\sigma)]$$

and let $\omega(t)$ be likewise continuously differentiable such that

$$\omega(t) > \max_i(\omega_i(t)), i = 1,2,3,4$$

for all t.

It is easy to show that

$$\mathcal{R}[\mathcal{D}^{aa}(t,t-\omega(t))] \supseteq \mathcal{R}[R(t,t-\omega(t)] \quad .$$

Hence there exists a matrix $K(t)$ such that

$$\mathcal{D}^{aa}(t,t-\omega(t))K(t) = R(t,t-\omega(t) \; .$$

By Corollary 10.1 it follows that $K(\cdot) \in C^1$.

Now define the $(n - r_c - r_{d_2} + r_{d_1}) \times (n - r_c - r_{d_2} + r_{d_1})$ matrix $T_4(t)$ by the formula

$$(10.14) \qquad T_4(t) = \begin{bmatrix} I_{r_{d_1}} & -K(t) \\ 0 & I_{n-r_c-r_{d_2}} \end{bmatrix}$$

where $I_{r_{d_1}}$ denotes the $r_{d_1} \times r_{d_1}$ identity matrix. Clearly, $T_4(t)$ is C^1. From (10.13) and (10.14) we obtain (omitting arguments on right hand side)

$$T_4'(t)\mathcal{D}_3(t,t-\omega(t))T_4(t) = \begin{bmatrix} D^{aa} & 0 \\ 0 & Q_1 \end{bmatrix}$$

where $Q_1 = Q - R'K$ is nonnegative definite and it follows by hypothesis that rank $Q_1(t,t - \omega(t)) = r_{d_3} - r_{d_1}$ for all t.

Applying Corollary 10.1 let $T_5(t)$ be an $(n - r_c - r_{d_2}) \times (n - r_c - r_{d_2})$ continuously differentiable nonsingular matrix such that

$$T_5'(t)Q_1(t,t - \omega(t))T_5(t) = \begin{matrix} P_1(t) & 0 \\ 0 & 0 \end{matrix}$$

where $P_1(t)$ is $(r_{d_3} - r_{d_1}) \times (r_{d_3} - r_{d_1})$ and rank $P_1(t) = r_{d_3} - r_{d_1}$ for all t. Then the coordinate transformation defined by

$$\begin{bmatrix} \tilde{x}^a(t) \\ \\ \tilde{x}^d(t) \end{bmatrix} = \hat{T}(t) \begin{bmatrix} x^a(t) \\ \\ x^d(t) \end{bmatrix},$$

where

$$\hat{T}(t) = \begin{bmatrix} I_{r_{d_1}} & 0 \\ \\ 0 & T_5(t)^{-1} \end{bmatrix} T_4(t)^{-1} = \begin{bmatrix} I_{r_{d_1}} & K \\ \\ 0 & T_5^{-1} \end{bmatrix} \tag{10.15}$$

has the effect of transforming the determinability matrix $\mathcal{D}_3$ for the system (10.11) into the form

$$\begin{bmatrix} D^{aa} & \vdots & \\ \cdots & \vdots & \cdots \\ & \vdots & \begin{bmatrix} P_1 & 0 \\ 0 & 0 \end{bmatrix} \end{bmatrix} \tag{10.16}$$

for all t. It is easy to check that $\hat{T}(t)^{-1}$ has the same form as $\hat{T}(t)$, in fact

$$\hat{T}(t)^{-1} = \begin{bmatrix} I_{r_{d_1}} & -KT_5 \\ 0 & T_5 \end{bmatrix} \tag{10.17}$$

for all t, where the dimensions of the "0" are $n - r_c - r_{d_2} \times r_{d_1}$. Therefore, the transformed state coefficient matrix of (10.11) is given by

$$\hat{A}_{a,d}(t) = \hat{T}(t)\begin{bmatrix} A^{aa}(t) & A^{ad}(t) \\ 0 & A^{dd}(t) \end{bmatrix}\hat{T}(t)^{-1} + \dot{T}(t)T(t)^{-1}$$

and has the form

$$\hat{A}_{a,d}(t) = \begin{bmatrix} \hat{A}^{aa}(t) & \hat{A}^{ad}(t) \\ 0 & \hat{A}^{dd}(t) \end{bmatrix}$$

and the corresponding transition matrix $\hat{\Phi}_{a,d}(t,\tau)$ also has the form, i.e.,

$$\hat{\Phi}_{a,d}(t,\tau) = \begin{bmatrix} \hat{\Phi}^{aa}(t,\tau) & \hat{\Phi}^{ad}(t,\tau) \\ 0 & \hat{\Phi}^{dd}(t,\tau) \end{bmatrix}$$

for all t,τ .

Now partition $\hat{\Phi}^{ad}$ and $\hat{\Phi}^{dd}$ as follows:

$$\hat{\Phi}^{ad} = \begin{bmatrix} \hat{\Phi}_1^{ad} & \hat{\Phi}_2^{ad} \end{bmatrix}$$

where $\hat{\Phi}_1^{ad}$ is $r_{d_1} \times n - r_c - r_{d_2} - r_{d_3} + r_{d_1}$, and

$$\hat{\Phi}^{dd} = \begin{bmatrix} \hat{\Phi}_{11}^{dd} & \hat{\Phi}_{12}^{dd} \\ \hat{\Phi}_{21}^{dd} & \hat{\Phi}_{22}^{dd} \end{bmatrix}$$

where $\hat{\Phi}_{12}^{dd}$ is $(n - r_c - r_{d_2} - r_{d_3} + r_{d_1}) \times (r_{d_3} - r_{d_1})$ and the remaining matrices are comformable with this.

This partition corresponds to a partition of the vector x^d as $\begin{bmatrix} x_1^d \\ x_2^d \end{bmatrix}$, where $\dim x_1^d = r_{d_3} - r_{d_1}$. Then the transpose of (10.18) becomes

$$\hat{\Phi}'_{a,d} = \begin{bmatrix} \Phi^{aa'} & 0 & 0 \\ \hat{\Phi}_1^{ad'} & \hat{\Phi}_{11}^{dd'} & \hat{\Phi}_{21}^{dd'} \\ \hat{\Phi}_2^{ad'} & \hat{\Phi}_{12}^{dd'} & \hat{\Phi}_{22}^{dd'} \end{bmatrix} \quad . \tag{10.19}$$

By (10.16), states which are determinable at any fixed time t must, under the new coordinate system, have the form

$$\begin{bmatrix} \hat{x}_1(t) \\ w(t) \end{bmatrix}$$

where $\dim \hat{x}_1(t) = r_{d_3}$, $\dim w(t) = n - r_c - r_{d_2} - r_{d_3} + r_{d_1}$, and $w(t) = 0$ for all t . From Theorem 7.6 and (10.19) it follows that

$$\begin{bmatrix} \hat{\Phi}_2^{ad'}(t) & \hat{\Phi}_{12}^{dd'}(t) \end{bmatrix} = \bigcirc$$

for all t . Transposing, and using equation (7.3), we find that

$$\hat{A}_{a,d} = \begin{bmatrix} A^{aa} & \hat{A}_1^{ad} & \bigcirc \\ 0 & \hat{A}_{11}^{dd} & \bigcirc \\ 0 & \hat{A}_{21}^{dd} & \hat{A}_{22}^{dd} \end{bmatrix}$$

for all t .

Before giving the final regrouping of terms, it remains to find the "output" coefficient matrix of (10.11) under the new coordinate system. This is given by

$$\hat{H}(t) = \begin{bmatrix} H^a(t) & 0 \end{bmatrix} \hat{T}(t)^{-1}$$

and so, from (10.17) (with arguments omitted)

$$\hat{H} = \begin{bmatrix} H^a & -H^a K T_5 \end{bmatrix} .$$

But from Theorem 10.3 it is clear that $\hat{H}$ takes on the form

$$\hat{H} = \begin{bmatrix} H^a & H_1^d & 0 \end{bmatrix}$$

where H_1^d is $r \times r_{d_3} - r_{d_1}$. We now define the following quantities:

$$\underset{\sim}{x}^a = x^a \qquad \underset{\sim}{x}^b = x^b$$

$$\underset{\sim}{x}^c = \begin{bmatrix} x^c \\ \\ x_1^d \end{bmatrix} \qquad \underset{\sim}{x}^d = x_2^d$$

$$\underset{\sim}{A}^{cc} = \begin{bmatrix} A^{cc} & \hat{A}_{11}^{dd} \end{bmatrix} \qquad \underset{\sim}{H}^c = \begin{bmatrix} H^c & H_1^d \end{bmatrix}$$

$$\underset{\sim}{A}^{dc} = \begin{bmatrix} A^{dc} & \hat{A}_1^{ad} \end{bmatrix} \qquad \underset{\sim}{A}^{dd} = \hat{A}_{22}^{dd}$$

$$\underset{\sim}{A}^{ac} = \begin{bmatrix} A^{ac} & \hat{A}_1^{ad} \end{bmatrix} \qquad \underset{\sim}{A}^{aa} = A^{aa} \quad \underset{\sim}{A}^{bb} = A^{ba}$$

and if we denote

$$A^{bd}\, x^d = \begin{bmatrix} A_1^{bd} & A_2^{bd} \end{bmatrix} \begin{bmatrix} x_1^d \\ \\ x_2^d \end{bmatrix}$$

then we define

$$\underset{\sim}{A}^{bc} = \begin{bmatrix} A^{bc} & A_1^{bd} \end{bmatrix} \qquad \underset{\sim}{A}^{bd} = \begin{bmatrix} A_2^{bd} \end{bmatrix} .$$

Finally, $\underset{\sim}{A}^{bb} = A^{bb}$, $\underset{\sim}{H}^a = H^a$, $\underset{\sim}{C}^a = C^a$, $\underset{\sim}{C}^b = C^b$. The theorem is thus proved.

It should be emphasized that our procedure for structural decomposition is "symmetric" from a number of points of view. For example, just as Theorem 10.3 is a dual result to Theorem 10.2, we could have given a completely dual procedure for obtaining a form consonant with (10.12). That is, one can easily write the dual to Theorems 10.4 and 10.5 which would begin with the application of Theorem 10.3 and would replace the matrices $\mathcal{D}_1$, $\mathcal{D}_2$, $\mathcal{D}_3$ with matrices $\mathcal{C}_1$, $\mathcal{C}_2$, $\mathcal{C}_3$ etc.

In addition to all this "dual" symmetry, the results of Section 7 indicate that the same type of structural decomposition is obtained if "controllability" is replaced by "reachability" and/or "determinability" is replaced by "observability". Hence, Theorems 10.4 and 10.5 as well as their duals are each representatives for a set four structural decomposition theorems.

To avoid confusion in the sequel, our discussion and interpretation of the results of this section are given only with reference to the actual procedure adopted in Theorems 10.4 and 10.5 to obtain (10.12). On the basis of our comments above, the reader can easily supply the interpretations for all the remaining approaches.

Remarks: 1. The overall coordinate transformation which produces the general structural decomposition of an arbitrary system (7.1) is represented by the matrix

$$T(t) = \begin{bmatrix} I_{r_{d_1}} & 0 \\ 0 & T_5(t)^{-1} \end{bmatrix} T_4(t)^{-1} \begin{bmatrix} T_2(t)^{-1} & 0 \\ 0 & T_3(t)^{-1} \end{bmatrix} T_1($$

2. For the special case when (1) is time-invariant, all applications of Corollary 10.1) will involve time-invariant transformations so that the procedure given in the proofs of Theorems 10.4 and 10.5 clearly leads to a time-invariant structural decomposition.

3. Pictorially, the decomposition (10.12) can be viewed as in Figure 1, which shows four interconnected systems S^a, S^b, S^c, S^d enclosed in "boxes" labelled with their associated state vectors. If, as is natural, we view the interconnecting lines inside the large "box" as input and output lines for the structural components, then the following result is readily discernible from individual examination of each structural component in (10.12) plus reference to the proofs of Theorems 10.4, 10.5.

Corollary 10.6:

(i) S^a is completely controllable and completely determinable

(ii) S^b is completely controllable and completely undeterminable

(iii) S^c is completely uncontrollable and completely determinable

(iv) S^d is completely uncontrollable and completely undeterminable

Note: If the matrices $A(\cdot)$, $C(\cdot)$, $H(\cdot)$ in (7.1) are analytic functions of time, the ranks of $\mathcal{C}(t,t+\mu(t))$, $\mathcal{D}_i(t,t-\omega_i))$, $i = 1, 2, 3$, will be constant everywhere in the t-domain. Hence, the system-theoretic interpretation of the structural decomposition of a system with analytic coefficients is given by Corollary 10.6. This provides the proof for assertions concerning the structural decomposition of analytic systems which were made by Kalman [16] and Weiss and Kalman [21]. It may be of interest to point out that in this special case the overall coordinate transformation can be taken to be analytic rather than just C^1. This follows directly from the proof of Dolezal's Theorem (Lemma 8.4) given by Weiss and Falb [26].

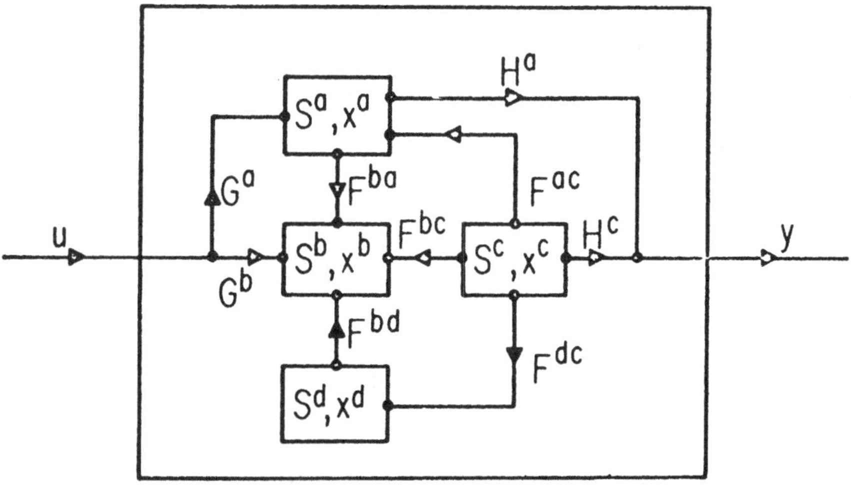

FIGURE 1: Structural Decomposition of a Linear System

11. WEIGHTING PATTERNS, IMPULSE RESPONSES, MINIMAL REALIZATIONS AND CONTROLLABILITY THEORY

Until recently, input/output relations were the most popular means used in engineering textbooks to represent systems, with the "State" being only implicitly considered. Since an input/output relation is the natural outcome of an attempt to model a system from experimental observations, it is of interest to investigate the relationship between input/output representations and state-vector representations of a system. In keeping with the theme of these notes, our discussion centers on the properties of controllability, observability, etc. for such representations. We consider only ordinary linear differential systems of the form (7.1).

The solution to (7.1) can be written as

$$y(t) = H(t)\Phi(t,t_o)x_o + \int_{t_o}^{t} W(t,\tau)u(\tau)d\tau \tag{11.1}$$

where x_o is the state of the system at time t_o and

$$W(t,\tau) = H(t)\Phi(t,\tau)C(\tau) \quad \text{for all} \quad t,\tau \tag{11.2}$$

and is denoted as the weighting pattern for (7.1) (See Weiss [20]). Clearly, if $x_o = 0$, then $W(t,\tau)$ contains all the information needed to compute all input/output pairs of the system. On this basis we concentrate our study of input/output relations solely on the function $W(t,\tau)$.

[A historical aside: Engineers have traditionally concerned themselves with input/output relations associated with the <u>causal impulse response function</u> $W_c(t,\tau)$ defined by

$$(11.3) \qquad W_c(t,\tau) = W(t,\tau) \;, \; t \geq \tau$$
$$= 0 \qquad , \; t < \tau \;.$$

The distinction between W_c and W from a system theoretic point of view, first discussed by Weiss and Kalman [21], is briefly considered later on].

The following definitions are pertinent to the discussion.

<u>Definition 11.1.</u> A weighting pattern (11.2) is in <u>reduced form</u> on an interval (t_1,t_2) if the rows of $\Phi(\eta,\cdot)C(\cdot)$ and the columns of $H(\cdot)\Phi(\cdot,\eta)$ are linearly independent functions on the interval (t_1,t_2) independent of η.

<u>Definition 11.2.</u> A weighting pattern of a system (7.1) is <u>globally reduced</u> if it is in reduced form on the entire interval $((-\infty,\infty))$ of definition of the system.

<u>Definition 11.3.</u> A weighting pattern has the <u>property</u> DCDO (for Differentially Controllable Differentially Observable) on an interval (t_1,t_2) if it is in reduced form on every subinterval of (t_1,t_2) of positive length.

Definition 11.4. The order of a globally reduced weighting pattern (11.2) is the number of columns of $H(\cdot)\Phi(\cdot,\eta)$ = number of rows of $\Phi(\eta,\cdot)C(\cdot)$.

Definition 11.5. A global realization of a globally reduced weighting pattern $W(t,\tau)$ is a dynamical system (7.1) whose weighting pattern can be reduced to $W(t,\tau)$.

Definition 11.6. If the dimension of the state space of the realization equals the order of $W(t,\tau)$, the realization is globally reduced or is minimal.

Perhaps the first thing to notice about $W(t,\tau)$ is that the problem of obtaining a global realization is trivial. This follows from the Lemma below.

Lemma 11.1. An $r \times p$ matrix function $W(t,\tau)$ is a weighting pattern for an n - dimensional system (7.1) if and only if W can be factored as

$$W(t,\tau) = \Psi(t)\Theta(\tau) \quad \text{for all} \quad t,\tau \tag{11.4}$$

where $\Psi(t)$ is $r \times n$, $\Theta(t)$ is $n \times p$, and $\Psi(\cdot)$, $\Theta(\cdot)$ are continuous functions.

Proof: (Sufficiency): The weighting pattern for the system $\{0,\Theta(\cdot),\Psi(\cdot)\}$ is (11.4).

(Necessity): Consider (11.1). Letting $\Psi(t) = H(t)\Phi(t,\eta)$ and $\Theta(t) = \Phi(\eta,t)C(t)$ yields the desired factorization.

It is also quite simple to construct a globally reduced weighting pattern from a given one, as indicated by the proof of the result below.

Lemma 11.2. Every weighting pattern has a globally reduced form.

Proof: Suppose (11.4) is not globally reduced. Then the rows of $\Theta(\cdot)$ and for the columns of $\Psi(\cdot)$ are dependent on $(-\infty,\infty)$. Suppose the rows of $\Theta(\cdot)$ are dependent. Then there exists an $n \times n$ nonsingular constant matrix K such that

$$K\Theta(t) = \begin{bmatrix} \hat{\Theta}(t) \\ 0 \end{bmatrix}$$

where the rows of $\hat{\Theta}(\cdot)$ are independent over $(-\infty,\infty)$. Then

$$W(t,\tau) = \Psi(t)K^{-1} \begin{bmatrix} \hat{\Theta}(t) \\ 0 \end{bmatrix} = \begin{bmatrix} \hat{\Psi}_1(t)\hat{\Psi}_2(t) \end{bmatrix} \begin{bmatrix} \hat{\Theta}(t) \\ 0 \end{bmatrix}$$

$$= \hat{\Psi}(t)\hat{\Theta}(t) .$$

If the columns of $\hat{\Psi}_1(t)$ are not independent over $(-\infty,\infty)$, we introduce a nonsingular constant matrix L such that

$$\hat{\Psi}_1(t)\, L = \begin{bmatrix} \tilde{\Psi}_{11}(t) & 0 \end{bmatrix}$$

where the columns of $\tilde{\Psi}_{11}(t)$ are independent over $(-\infty,\infty)$.

Then, letting $L^{-1}\hat{\Theta}(t) = \tilde{\Theta}(t)$ we get

$$W(t,\tau) = \tilde{\Psi}_{11}(t)\tilde{\Theta}(\tau) \quad \text{for all} \quad t,\tau \tag{11.5}$$

and (11.5) is globally reduced.

We now investigate some of the properties of minimal realizations of globally reduced weighting patterns. The first result justifies the terminology "minimal".

Lemma 11.3. A minimal realization of a globally reduced weighting pattern $W(t,\tau)$ has the lowest dimension of all global realizations of $W(t,\tau)$.

Proof. Suppose the contrary. Let n be the order of $W(t,\tau)$ and consider a global realization of $W(t,\tau)$ with dimension $< n$. Clearly, its weighting pattern is of order $< n$ which contradicts the assumption that $W(t,\tau)$ is globally reduced.

An interesting fact which relates the material on "structure" to that presently being developed is the following. (See Kalman [15], [16]).

Proposition 11.4. The subsystem S^a in Figure 1 is a minimal realization of the weighting pattern for the overall system.

Proof: The weighting pattern for the system (10.12) is

$$W(t,\tau) = H^a(t)\Phi^{aa}(t,\tau)C^a(\tau) \quad \text{for all} \quad t,\tau$$

where Φ^{aa} corresponds to the coefficient matrix F^{aa} . The right side of (11.6) is the weighting pattern for S^a and is globally reduced. It is a trivial matter to check that the order of this weighting pattern is the dimension of $\underset{\sim}{X}^a$.

Definition 11.7. Two linear systems $S,\hat{S}$ of the form (7.1) are algebraically equivalent if there exists a nonsingular continuously differentiable matrix $T(t)$ such that

$$A_{\hat{s}}(t) = T(t)A_s(t)T(t)^{-1} + \dot{T}(t)T(t)^{-1}$$

(11.7) $$C_{\hat{s}}(t) = T(t)C_s(t)$$

$$H_{\hat{s}}(t) = H_s(t)T(t)^{-1}$$

for all t .

An obvious result is

Proposition 11.5. Weighting patternsare invariant under algebraic equivalence.

Lemma 11.6. Points of time from which a system (7.1) is controllable (or observable) or at which a system is reachable (or determinable) are invariant under algebraic equivalence.

Proof: (for controllability only. The rest follows analogously). Under algebraic equivalence we have the correspondence

$$C(t,t + \mu(t) \to T(t)C(t,t + \mu(t))T'(t) = \hat{C}(t,t + \mu(t))$$

and so rank $C(t,t + \mu(t)) = n$ implies rank $\hat{C}(t,t + \mu(t)) = n$.

The following result was first stated but not proved in [16] and [21]. A proof was subsequently published by Youla [26]. The following proof combines that of Youla with one given by Kalman in unpublished notes.

Theorem 11.7. Any two minimal realizations of a given globally reduced weighting pattern are algebraically equivalent.

Proof: It is clear from the proof of Lemma 11.1 that any minimal realization is algebraically equivalent to one with the coefficient matrix $A(\cdot) \equiv 0$ (take $T(t) = \Phi(t,\eta)$ in (11.7)). Hence it suffices to prove that any two minimal realizations $\{0,\Psi_1(\cdot),\Theta_1(\cdot)\}$ and $\{0,\Psi_2(\cdot),\Theta_2(\cdot)\}$ of a given globally reduced weighting pattern $W(t,\tau)$ are algebraically equivalent. To do this, first note that

$$W(t,\tau) = \Psi_1(t)\Theta_1(\tau) = \Psi_2(t)\Theta_2(\tau) \tag{11.8}$$

where the columns of the $\Psi_i(\cdot)$ and the rows of the $\Theta_i(\cdot)$, $i = 1,2$ are linearly independent on $(-\infty,\infty)$. It then follows that there exist finite intervals J_i, K_i, $i = 1,2$ on which the aforementioned columns and rows are linearly independent respectively. [for if not, then on any interval $[-k,k]$, $k = 1,2,\ldots,$ there exists a constant vector ξ_k of unit norm such that $\Psi_1(t)\xi_k = 0$ almost everywhere on $[-k,k]$. The sequence $\{\xi_k\}$ has a convergence subsequence $\{\xi_{k_i}\}$ such that $\lim_{i\to\infty} \xi_{k_i} = \xi$ and $\Psi_1(t)\xi = 0$ a.e. in $(-\infty,\infty)$ thus contradicting the linear independence of the columns of $\Psi(\cdot)$. A similar argument holds for $\Psi_2(\cdot)$, $\Theta_1(\cdot)$, $\Theta_2(\cdot)]$. Hence the matrices

$$M_i = \int_{J_i} \Psi_i'(t)\Psi_i(t)dt \ , \ i = 1,2$$

and

$$N_i = \int_{K_i} \Theta_i(t)\Theta_i'(t)dt \ , \ i = 1,2$$

are nonsingular. Hence, from (11.8) we can write that

$$\Theta_2(\tau) = M_2^{-1} \int_{J_2} \Psi_2'(t)\Psi_1(t)dt \ \Theta_1(\tau) = U\Theta_1(\tau) \tag{11.9}$$

$$\Psi_2(t) = \Psi_1(t) \int_{K_2} \Theta_1(\tau)\Theta_2'(\tau)d\tau \ N_2^{-1} = \Psi_1(t)V \tag{11.10}$$

Now, multiplying both sides of (11.10) on the left by $\Psi_2'(t)$, integrating over J_2 , and multiplying on the left again by M_2^{-1} we get $I_n = UV$. But from (11.8), (11.9), (11.10), we have

$$\Psi_1(t)\Theta_1(\tau) = \Psi_1(t)VU\Theta_1(\tau)$$

from which it follows that $I_n = VU$ so that $U = V^{-1}$ and therefore $\{0,\Psi_1(\cdot),\Theta_1(\cdot)\}$ is algebraically equivalent to $\{0,\Psi_2(\cdot),\Theta_2(\cdot)\}$ which proves the theorem.

As a direct consequence of Theorem 11.7 and Lemma 11.6 we have the statement that <u>all minimal realizations of a given globally reduced weighting pattern have essentially the same behavior with respect to the properties of controllability, reachability, determinability, and observability</u>.

We can, of course, go even further as indicated below.

<u>Theorem 11.8.</u> Given a globally reduced weighting pattern (11.1), there exist finite values of time t', t'', such that all minimal realizations of (11.1) are controllable (or observable) from all $t < t'$ and are reachable (or determinable) at all $t > t''$.

<u>Proof:</u> Let $W(t,\tau) = \Psi(t)\Theta(\tau)$ where $\Psi(t) = H(t)\Phi(t,\eta)$, $\Theta(\tau) = \Phi(\eta,\tau)G(\tau)$. Since the rows of $\Theta(\cdot)$ (columns of $\Psi(\cdot)$) are linearly independent over $(-\infty,\infty)$, by the argument used in the proof of Theorem 11.7, there exists a finite value of time $t'(t'')$ such that rows of $\Theta(\cdot)$ (columns of $\Psi(\cdot)$) are linearly independent over $[t',\infty)$ $((-\infty,t''])$. The rest follows from Lemma 11.6.

<u>Corollary 11.9.</u> All minimal realizations of a weighting pattern with the property DCDO are differentially controllable, reachable, determinable, observable.

We now relate the concept of <u>impulse response</u> to the material developed thus far.

It used to be the standard practice in engineering textbooks (especially those concerned with communication theory) to represent linear systems by an input/output expression of the form

$$y(t) = \int_{-\infty}^{\infty} W_c(t,\tau)u(\tau)d\tau \tag{11.11}$$

(The infinite limits apparently motivated by the heavy use of Fourier transforms in communications problems). The function $W_c(t,\tau)$ was normally referred to as the "impulse response" of the system since if one replaces $u(\tau)$ by a vector whose i^{th} component is a dirac δ - function $\delta(\tau - \zeta)$, then by the formal properties of δ - functions, $y(t) \rightarrow$ the ith row of $W_c(t,\xi)$. Having thus introduced the viewpoint of the physicist into the problem, it was a simple matter to note that since "physical" systems do not produce responses prior to introduction of stimuli, one should assume that for physical or <u>causal</u> systems, the function $W_c(t,\tau)$ in (11.11) has the property that $W_c(t,\tau) = 0$ for $t < \tau$. It was perhaps inevitable that this widely taught example of model building via the incoproration of physical principles into an a priori assumed mathematical form (11.11) would be (incorrectly, as it turned out) associated with the mathematical properties of solutions to ordinary differential equations.

Nevertheless there is a problem involved for, in fact, we can obtain information about the weighting pattern $W(t,\tau)$ through experiment only for $t \geqslant \tau$. It is therefore of some interest to rephrase the results obtained thus far so as to apply to "impulse responses". The causal impulse response function was defined by (11.3). Of interest also is the anticausal impulse response function defined as follows.

Definition 11.8. The anticausal impulse response $W_a(t,\tau)$ of a system (7.1) with weighting pattern $W(t,\tau)$ is defined by

$$
(11.12) \qquad \begin{aligned} W_a(t,\tau) &= W(t,\tau) \ , \ t \leqslant \tau \\ &= \quad 0 \ , \ t > \tau \ . \end{aligned}
$$

Definition 11.9. A realization of a causal (anticausal) impulse response $W_c(t,\tau)$ $(W_a(t,\tau))$ is a system (7.1) whose causal (anticausal) impulse response is $W_c(t,\tau)$ $(W_a(t,\tau))$.

The following Corollary of Lemma 11.1 is obvious.

Corollary 11.10. An $r \times p$ matrix function $W_c(t,\tau)$ is a causal impulse response for an n - dimensional system (7.1) if and only if there exists an $r \times n$ matrix $\Phi(\cdot)$, and an $n \times p$ matrix $\Theta(\cdot)$, both defined and continuous for all time such that

$$
(11.13) \qquad \begin{aligned} W_c(t,\tau) &= \Psi(t)\Theta(\tau) \ , \ t \geqslant \tau \\ &= \quad 0 \ , \ t < \tau \ . \end{aligned}
$$

In similar fashion, the anticausal impulse response must have the form

(11.14) $$W_a(t,\tau) = \Psi(t)\Theta(\tau) \ , \ t \leq \tau$$
$$= \qquad 0 \ , \ t > \tau \ .$$

Definition 11.10. A causal impulse response (11.13) is globally reduced if for some $\tau > -\infty$, the rows of $\Theta(\cdot)$ are linearly independent over $(-\infty,\tau]$ while the columns of $\Psi(\cdot)$ are linearly independent over $[\tau,\infty)$.

Definition 11.11. An anticausal impulse response (11.14) is globally reduced if, for some $\xi < \infty$, the rows of $\Theta(\cdot)$ are linearly independent over $[\xi,\infty)$ while the columns of $\Psi(\cdot)$ are linearly independent over $(-\infty,\xi]$.

Definition 11.12. Given a globally reduced causal (anticausal) impulse response (11.13)((11.14)), the function $W(t,\tau) = \Psi(t)\Theta(\tau)$ defined for all t,τ is the naturally induced weighting pattern associated with that impulse response. (Note that this weighting pattern must be globally reduced).

Remark: It should be emphasized that many different weighting patterns may be associated with the graph of a given (causal or anticausal) impulse response. For example, let $\Theta(\tau) = 0$ for all $\tau \leq 0$ in (11.13). Then $\Psi(t)$ can be arbitrarily chosen for $t \leq 0$ without affecting $W_c(t,\tau)$. Definition 11.11

simply identifies the weighting pattern resulting from a choice of $\Psi(\cdot)$ and $\Theta(\cdot)$ from the class of possible choices.

Definition 11.13. A realization (7 .1) of a globally reduced impulse response is minimal if its dimension equals the order of the weighting pattern naturally induced by that impulse response.

It is obvious from the previous Remark that two minimal realizations of a given globally reduced impulse response may not be algebraically equivalent. One can say the following, however, as a result of Theorem 11.7.

Theorem 11.11. Two minimal realizations of a globally reduced causal (anticausal) impulse response are algebraically equivalent if and only if they have the same anticausal (causal) impulse response.

Our final result touches on the question of what information regarding control-theoretic properties of a system is carried by the system's impulse response, and follows from Definitions 11.10, 11.11, plus earlier results.

Let τ,ξ be as in Definitions 11.10, 11.11. Then we have

Theorem 11.12. (a) A minimal realization of a globally reduced causal impulse response is reachable at all $t > \tau$ and is observable from all $t < \tau$.

(b) A minimal realization of a globally reduced anticausal impulse response is controllable from all $t < \xi$ and is determinable at all $t > \xi$.

BIBLOGRAPHY

[1] E. G. Al'brekht and N. N. Krasovskii, "The Observability of a Nonlinear Controlled System in the Neighborhood of a given Motion", Automation and Remote Control, 25, pp. 934-944, (1965).

[2] R. Bellman and K. L. Cooke, Differential-Difference Equations, Academic Press, New York, (1963).

[3] C. Carathéodory, "Untersuchunger Über die Grundlagen der Thermodynamik", Math. Ann. pp. 355-386, (1909).

[4] W. L. Chow, "Über Systeme von Linear Partiallen Differentialgleichungen erster Ordnung", Math. Ann., pp. 95-105, (1940).

[5] D. H. Chyung and E. B. Lee, "Optimal Systems with Time-Delays", Proc. 3rd Congress of the IFAC, London, (1966).

[6] E. A. Coddington and N. Levinson, Theory of Ordinary Differential Equations, McGraw-Hill, New York, (1955).

[7] E. Davison, L. Silverman and P. Varaiya, "Controllability of a Class of Nonlinear Time-Variable Systems", IEEE Trans. Auto. Control, AC-12, pp. 791-792, (1967).

[8] J. Dieudonné, Foundations of Modern Analysis, Academic Press, New York, (1960).

[9] V. Doležal, "The Existence of a Continuous Basis of a Certain Linear Subspace of E_r which Depends on a Parameter", Cas. Pro. Pest. Mat., 89, pp. 466-468, (1964).

[10] R. Driver, "Existence and Stability of Solutions of a Delay-Differential System", Arch. Rat. Mech. Anal., 10, pp. 401-426, (1962).

[11] E. G. Gilbert, "Controllability and Observability in Multivariable Control Systems", J. SIAM Control, 1, pp. 128-151, (1953).

[12] J. K. Hale and K. R. Meyer, "A Class of Functional Equations of Neutral Type", Mem. Amer. Math. Soc., No. 76, Providence, (1967).

[13] H. Hermes, "Controllability and the Singular Problem", J. SIAM Control, 3, pp. 241-260, (1965).

[14] R. E. Kalman, "On the General Theory of Control Systems", Proc. First IFAC, Butterworth's, London, pp. 481-492 (1960).

[15] R. E. Kalman, Y. C. Yo and K. S. Narendra, "Controllability of Linear Dynamical Systems", Contr. to Diff. Eqs., 1, pp. 189-213, (1962).

[16] R. E. Kalman, "Mathematical Description of Linear Dynamical Systems", J. SIAM Control, 1, pp. 152-192, (1963).

[17] R. E. Kalman, "Contributions to the Theory of Optimal Control", Bol. Soc. Mat. Mex., pp. 102-119, (1960).

[18] F. M. Kirillova and S. V. Čurakov, "On the Problem of Controllability of Linear Systems with Aftereffect" (Russian), Diff. Urav., 3, pp. 436-445, (1967).

[19] E. B. Lee and L. Markus, "Optimal Control for Nonlinear Processes", Arch. Rat. Mech. Anal., 8, pp. 36-38, (1961).

[20] L. Weiss and R. E. Kalman, "Contributions to Linear System Theory", Int'l J. Engrg. Sci., 3, pp. 141-171 (1965).

[21] L. Weiss, "Weighting Patterns and the Controllability and Observability of Linear Systems", Proc. Nat. Acad. Sci. (USA), 51, pp. 1122-1127, (1964).

[22] L. Weiss, "The Concepts of Differential Controllability and Differential Observability", J. Math. Anal. and Appl., 10, pp. 442-449 (1965). "Correction and Addendum", J. Math. Anal. and Appl., 13, pp. 577-578, (1966).

[23] L. Weiss, "On the Controllability of Delay-Differential Systems", J. SIAM Control, 5, pp. 575-587, (1967).

[24] L. Weiss, "On the Structure Theory of Linear Differential Systems", J. SIAM Control, (1968).

[25] L. Weiss and P. L. Falb, "Doležal's Theorem, Linea Algebra with Continuously Parametrized Elements and Time-Varyin Systems", Math. Systems Theory, (1969).

[26] D. C. Youla, "The Synthesis of Linear Dynamical Systems from Prescribed Weighting Patterns", J. SIAM Appl. Math., 14, pp. 527-549, (1966).

OTHER PERTINENT REFERENCES

27. H. A. Antosiewicz, "Linear Control Systems", Arch. Rat. Mech. Anal., 12, pp. 313-324 (1963).

28. A. V. Balakrishnan, "On the Controllability of Nonlinear System", Proc. Nat. Acad. Sci. (USA), 55, pp. 465-468, (1966).

29. R. Conti, "Contributions to Linear Control Theory", J. Diff. Eqs., 1, pp. 427-445, (1965).

30. C. A. Desoer and P. P. Varaiya, "Minimal Realizations of Nonanticipative Impulse Responses", J. SIAM Appl. Math., 15, pp. 754-764, (1967).

31. H. O. Fattorini, "On Complete Controllability of Linear Systems", J. Diff. Eqs.,

32. A. Halanay, Differential Equations: Stability, Oscillations, Time Lags, Academic Press, New York, 1966.

33. B. L. Ho and R. E. Kalman, "Effective Construction of State-Variable Models from Input/Output Functions", Regelungstechnik, 12, pp. 545-548, (1966).

34. E. Kreindler and P. Sarachik, "On the Concepts of Controllability and Observability of Linear Systems", IEEE Trans Auto. Control, AC-9, pp. 129-136, (1964).

35. N. N. Krasovskii, "On the Theory of Controllability and Observability of Linear Dynamic Systems", PMM, 28, pp. 3-14, (1964).

36. C. E. Langenhop, "On the Stabilization of Linear Systems", Proc. Amer. Math. Soc., 15, pp. 735-742, (1964).

37. J. P. LaSalle, "The Time-Optimal Control Problem", Contributions to the Theory of Nonlinear Oscillations, Vol. V, Princeton Univ. Press, pp. 1-24, (1960).

38. L. Markus, "Controllability of Nonlinear Processes", J. SIAM Control, 3, pp. 78-90, (1965).

39. L. S. Pontryagin, V. G. Boltyanskii, R. V. Gamkrelidze and E. F. Mishchenko, The Mathematical Theory of Optimal Processes, Wiley (Interscience), New York, (1962).

40. D. L. Russell, "Nonharmonic Fourier Series in the Control Theory of Distributed Parameter Systems", J. Math. Anal. Appl., 18, pp. 542-560, (1967).

41. L. Silverman and H, Meadows, "Controllability and Observability in Time-Variable Linear Systems", J. SIAM Control, 5, pp. 64-73, (1967).

42. W. M. Wonham, "Pole Assignment in Multi-Input Controllable Linear Systems", IEEE Trans. Auto. Control, AC-12, pp. 660-665, (1967).